THE ORIGIN OF CHONDRULES AND CHONDRITES

Chondrites are the largest group of meteorites. They can provide unique insights into the origins and early evolution of our Solar System, and even into the relationships between our Solar System and other stars in the vicinity of our Sun. The largest structural components of most chondrites are the glass-bearing chondrules, and there are numerous theories for their origin. This clear and systematic text summarizes the ideas surrounding the origin and history of chondrules and chondrites, drawing on research from the various scientific disciplines involved. With citations to every known published paper on the topic, it forms a comprehensive bibliography of the latest research, and extensive illustrations provide a clear visual representation of the scientific theories. This text will be a valuable reference for graduate students and researchers in planetary science, geology, and astronomy.

DEREK SEARS was born in England and obtained a bachelor's degree in chemistry at the University of Kent at Canterbury, and a Ph.D. in Astronomy and Geology at the University of Leicester. He is now Professor of Chemistry and Director of the Arkansas–Oklahoma Center for Space and Planetary Sciences. He teaches chemistry and performs meteorite research, and is currently involved in creating new research and graduate teaching programs in space and planetary sciences. Professor Sears is probably best known for his pioneering studies on the use of thermoluminescence to characterize primitive meteorites and to determine the thermal and radiation history of Antarctic meteorites. In 1999 he received the University of Arkansas' highest award for research and service, and asteroid 4473 Sears was named in his honor. This is his third book on meteorites.

CAMBRIDGE PLANETARY SCIENCE

Series Editors: Fran Bagenal, David Jewitt, Carl Murray, Jim Bell, Ralph Lorenz, Francis Nimmo, Sara Russell

Books in the series

1. *Jupiter: The Planet, Satellites and Magnetosphere*
   Edited by Fran Bagenal, Timothy E. Dowling, and William B. McKinnon
   0 521 81808 7
2. *Meteorites: A Petrologic, Chemical, and Isotopic Synthesis*
   Robert Hutchison
   0 521 47010 2
3. *The Origin of Chondrules and Chondrites*
   Derek W. G. Sears
   0 521 83603 4

THE ORIGIN OF CHONDRULES AND CHONDRITES

DEREK W. G. SEARS
*Arkansas–Oklahoma Center for Space and Planetary Sciences,
and Department of Chemistry and Biochemistry,
University of Arkansas, USA*

PUBLISHED BY THE PRESS SYNDICATE OF THE UNIVERSITY OF CAMBRIDGE
The Pitt Building, Trumpington Street, Cambridge, United Kingdom

CAMBRIDGE UNIVERSITY PRESS
The Edinburgh Building, Cambridge, CB2 2RU, UK
40 West 20th Street, New York, NY 10011–4211, USA
477 Williamstown Road, Port Melbourne, VIC 3207, Australia
Ruiz de Alarcón 13, 28014 Madrid, Spain
Dock House, The Waterfront, Cape Town 8001, South Africa
http://www.cambridge.org

© D. Sears 2004

This book is in copyright. Subject to statutory exception
and to the provisions of relevant collective licensing agreements,
no reproduction of any part may take place without
the written permission of Cambridge University Press.

First published 2004

Printed in the United Kingdom at the University Press, Cambridge

Typeface Times 11/14 pt.   *System* LATEX 2_ε [TB]

A catalog record for this book is available from the British Library

Library of Congress Cataloging in Publication data

ISBN 0 521 83603 4 hardback

The publisher has used its best endeavors to ensure that the URLs for external websites referred to in this book
are correct and active at the time of going to press. However, the publisher has no responsibility for the websites
and can make no guarantee that a site will remain live or that the content is or will remain appropriate.

Contents

List of figures	*page* vii
List of tables	ix
Preface	xi
1 Historical introduction	1
1.1 Rocks from the sky	1
1.2 Museums and collectors	4
1.3 The instruments	4
1.4 The space age	10
1.5 The great expeditions	11
1.6 Cosmic sediments	12
2 Potential meteorite parent bodies	21
2.1 Asteroids as potential meteorite parent bodies	21
2.2 Impact and cratering processes	40
3 Chondrites and their main properties	49
3.1 Classification and composition	49
3.2 Formation history	60
3.3 The challenge	70
4 Chondrules and their main properties	73
4.1 The diversity of chondrules	73
4.2 Chondrule groups	75
4.3 Composition of chondrules	80
4.4 Physical processes affecting chondrule history	90
4.5 Chondrule rims and matrix – implications for formation history	94
4.6 Stable isotope studies of chondrules	98
4.7 Radiogenic isotope studies of chondrules	103
4.8 Interclass comparisons	105
4.9 Refractory inclusions	106

	4.10 Relationship between chondrules and refractory inclusions	107
	4.11 "Chondrules" from other planetary bodies	109
5	Theories for the origin of chondrules	111
	5.1 Some general comments	111
	5.2 Processes occurring in the primordial solar nebula	112
	5.3 Processes occurring on parent bodies	119
6	Discussion of theories for the origin of chondrules	125
	6.1 The primordial solar nebula and possible cosmochemistry	125
	6.2 Critique of nebula theories for chondrule formation	129
	6.3 Formation of chondrules by impact into a regolith	137
7	Discussion of theories for metal–silicate fractionation	141
	7.1 Chondrule sorting	141
	7.2 The metal–silicate fraction in the nebula	143
	7.3 Metal–silicate fractionation on the parent body	147
8	So how far have we come and where do we go next?	151
	8.1 Chondrules and chondrite classes as impact pyroclastics	151
	8.2 The details	154
	8.3 So far, so near	158
	8.4 Why the impasse?	159
	8.5 Breaking the log jam?	160
	References	163
	Index	199

Figures

1.1	Edward Howard – an early investigator	*page* 2
1.2	Cartoon showing texture and components of chondrites	3
1.3	Chondrites under the microscope	5
1.4	Solar vs chondrite composition	7
1.5	Urey–Craig plot	8
1.6	The age of chondrites	9
1.7	Astrophysical history of chondritic material	13
1.8	Mineral and phase composition of chondrites	16
1.9	Chondrule diameters vs metal grain sizes	16
1.10	Metamorphism in chondrites	17
1.11	Brecciated textures of chondrites	19
2.1	Orbital parameters for asteroids	24
2.2	Spectra of asteroids	26
2.3	Distribution of asteroid and meteorite classes	28
2.4	Texture of gas-rich regolith breccias	30
2.5	Cosmic ray exposure ages for chondrites	33
2.6	Ar–Ar ages for chondrites	34
2.7	Thermal models for chondrites	36
2.8	Cartoon of hypothetical ejecta blanket	37
2.9	Structure of a meteorite parent body	38
2.10	Asteroids and their surfaces	41
2.11	Fraction of ejecta escaping	44
2.12	Melt produced by asteroid impact	46
3.1	Phase diagrams for Mg–Fe silicates in a solar gas	52
3.2	Composition of chondrites	54
3.3	Thermodynamic calculations for a solar gas	56
3.4	Three oxygen isotope plot for chondrites	59
3.5	Petrographic evidence for aqueous alteration	65
3.6	Initial $^{87}Sr/^{86}Sr$ ratios in chondrites	66
3.7	I–Xe ages of chondrites	68
4.1	Representative chondrule textures (Tschermak)	74
4.2	Representative chondrule textures (Merrill)	76

4.3	Representative chondrule textures (Scott and Taylor)	78
4.4	Definition of compositional chondrule classes	80
4.5	Bulk composition of chondrules	83
4.6	Factor analysis of chondrule compositions	84
4.7	Laboratory experiments on composition of chondrules	86
4.8	Proposed phase diagrams for chondrule and matrix formation	87
4.9	Chondrules as open and closed systems	89
4.10	Production of synthetic chondrules	91
4.11	Formation of chondrule rims	95
4.12	Chondrule diameter vs rim thickness	96
4.13	Oxygen isotopes in chondrules	99
4.14	Oxygen isotopes in a zoned chondrule	100
4.15	Oxygen isotopes in FUN inclusions	102
4.16	Oxygen isotopes in an Allende inclusion	103
4.17	Xenon isotopes in chondrules	104
4.18	Chondrules from non-asteroid parent bodies	108
5.1	Chondrules from the nebula	113
5.2	Cartoon – chondrules from parent bodies	120
6.1	Formation of the Solar System	126
6.2	Stability of minerals and the adiabat	128
6.3	Fragmentation of meteorites in the atmosphere	132
6.4	Charged particle tracks and microcraters	136
6.5	Outcomes for ejecta impacting an asteroid	139
7.1	Schematic drawings of metal–silicate separation in nebula	144
7.2	Aerodynamic sorting of chondrules and metal	146
7.3	Cartoon – metal–silicate separation on a parent body	147
8.1	Mount Saint Helens	153
8.2	Pyroclastic flows	154
8.3	Cartoon – the history of a chondrule	155
8.4	Cartoon – the history of chondrites	156
8.5	The Hayabusa spacecraft	161
8.6	The Hera spacecraft	162

Tables

1.1	The chondrite classes	*page* 14
1.2	The petrographic types	18
2.1	Asteroid densities	22
2.2	Water and density of C chondrites	23
2.3	Asteroids and water	23
2.4	The asteroid–meteorite link	27
2.5	Asteroids visited by spacecraft	40
2.6	Regoliths on asteroids	45
3.1	Temperatures and cooling rates for chondrites	62
3.2	Aqueous alteration of CM chondrites	64
4.1	Compositional groups of chondrules	79
4.2	References to analyses of chondrules	81
4.3	Comparison of chondrule classification schemes	82
4.4	Open- and closed-system evolution of chondrules	90
4.5	Temperatures and cooling rates for chondrules	93
4.6	Chondrule classes and chondrite classes	105
7.1	Fluidization and metal–silicate ratios	148

Preface

Rocks falling from the sky have a long and colorful history. I mean this both in a socio-economic sense and, perhaps more obviously, in a scientific sense. Stories of stones from the heavens have been with us for as long as humans have left traces of themselves. In ancient tombs and burial sites, in their earliest writings and during the faltering steps of the industrial revolution and the creation of modern science, people wrote about rocks from the sky now known as meteorites. In many respects the history of modern science instrumention is inextricably linked with the history meteorite studies.

Meteorites are major witnesses of the history of our Solar System. Everyone agrees that meteorites are ancient materials from the earliest stages in the history of the Solar System. Their age, composition, and texture clearly point to this conclusion. Everyone also agrees that meteorites are fragments from near-Earth asteroids, which occasionally threaten us with impact, and it seems that such asteroids largely come out of the Main Asteroid Belt between Mars and Jupiter although a small fraction of them are probably related to comets. These rocks are fascinating to study. They are sufficiently like terrestrial rocks that similar techniques and approaches can be used, yet they present a whole new range of physical and chemical processes to consider, processes that take the researcher from petrologist, mineralogist, and geochemist to the astronomer and the astrophysicist. But while they reward us with many new observations and insights, much about them remains covered in a veil of obscurity "of truly delphic proportions." For example, what is the origin of the chondrules from which chondrites get their name? What processes have given rise to the differences in the accumulation of metal and silicates that characterize the various classes?

This book emerged from a paper I was invited to give at the annual conference on Antarctic meteorites hosted by the National Institute for Polar Research (NIPR) in Tokyo. I am very grateful to K. Yanai and H. Kojima for the invitation and their extraordinary hospitality. In an age of endless specialization and highly focussed expertise, I wanted to present a discussion of the big picture – laying out the variety of ideas that have been published and trying to stimulate some new thoughts. I wanted to give an overview of both where we have been in our thinking and where we are now, whilst remaining very aware that many major issues in the study of these precious rocks have not yet been resolved. I also wanted to do this in an easily digestible form. So throughout the book appear lists of theories, cartoons, and figures. Lists can be dull, but they can be read easily, used for reference, and they give an idea of real constraints that exist on some of our theories. I also wanted to collect together in one place as many literature references as possible, because many good ideas are becoming lost in the explosion in recent literature. I wonder how many of our new ideas are restatements of old ideas and I wonder how many good ideas were prematurely interred.

In addition to my NIPR hosts, I am grateful to a number of people for helping me assemble this book. Simon Mitton of Cambridge University Press persuaded me to finish what had become a decade-long project. Four anonymous reviewers gave me an objective perspective on what I was proposing to do that encouraged me to finish and helped me improve the project. The University of Arkansas has provided the means for me to achieve much that I have done, including this book. Hazel Sears helped in the mechanics of book assembly and proofed the final product. To them all my thanks, and I hope they feel I have justified their efforts.

1
Historical introduction

1.1 Rocks from the sky

The ancients observed and collected rocks that fell from the sky. There are reports of the Romans, Greeks, Egyptians, Japanese, and the natives of North America and other countries collecting them, using them for trade, and putting them in places of importance such as tombs. Modern-age research on such objects opened with the pioneering work of Howard and Bournon (Fig. 1.1). Aristocrats Edward Charles Howard and Jacques Louis Compte de Bournon published arguably the first scientific investigation of these rocks or meteorites as they are now known (Howard, 1802). They found that these rocks, regardless of where in the world they fell, contained metal, sulfide, stony materials, and very often "curious globules" in varying amounts. What was also remarkable was that the metal in all these rocks from the sky – whether it was tiny metallic grains in the more stony meteorites or the large masses of "native iron," now known to be iron meteorites – was found to contain nickel, a novel and only recently discovered element. Clearly, from the first day that these rocks were seriously examined their major components, which would have to be explained, were identified. As the nineteenth century unfolded, Greek names were attached to these objects, and the globules became "chondrules" and the type of rocks that contain them became "chondrites."

Metal, sulfide, and stony materials had been seen before, but these curious globules and nickel-bearing metal were unique. Fig. 1.2 is a visual summary of the major components in chondrites as currently known: metal, sulfide, matrix, refractory inclusions (calcium–aluminum-rich inclusions, CAI), and chondrules with a variety of internal textures, some coated with rims of matrix-like material.

In the latter part of the eighteenth century there was an almost universal belief, following a report by Lavoisier and others, that meteorites were produced by lightning (Fougeroux et al., 1772; Lavoisier, 1772). The burned outer surface of the meteorite might point to this conclusion, but one suspects that Lavoisier's encounter with a

Figure 1.1 Edward Howard (center) as he appears in an engraving at the Royal Institution of London. The engraving is a copy of a profile in bronze, probably made posthumously. Howard, with petrologist Jacques Louis Compte de Bournon, arguably performed the first modern study of meteorites. His report – which included the first observation of curious globules, i.e. chondrules, in meteorites – was critical in triggering serious scientific study of meteorites.

roof tile dislodged by a lightning stroke was a bigger factor. Certainly, Lavoisier's colleagues were not as attracted to the conclusion that meteorites were terrestrial rocks struck by lightning as was Lavoiser. However, between the publication of Chladni's book in 1794 and Howard's paper of 1802 there arose a widespread acceptance that meteorites actually fell from the sky and some even believed that the stones had an extraterrestrial origin. I think that this was primarily a consequence of the early chemical and physical work that revealed the similarity of meteorites to each other, regardless of country of fall, and their dissimilarity to local country rocks (Sears, 1976). However, other writers have argued that the large number of falls at the time (Burke, 1986) and Chladni's eloquence (Marvin, 1996) were primarily responsible for the swing of opinion. In any event, by 1803

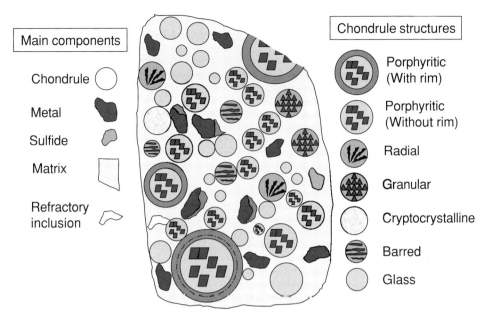

Figure 1.2 Cartoon showing the variety of components in a typical primitive (low-petrographic type) ordinary chondrite. The components are chondrules, metal, sulfide, matrix, and refractory inclusions. The term "matrix" is used for a variety of materials in meteorites. In this instance we mean the very fine-grained, rim-like matrix seen only in the most primitive meteorites. The dimensions of this hypothetical section would be about 1 cm × 2 cm.

some of the world's leading scientists were engaged in studies of these rocks from space.

From 1802 to about 1840 the most widely accepted theory for the origin of meteorites was that they came from the Moon (Olbers, 1803; Poisson, 1803; Berzelius, 1834), but there were also proponents of theories involving a terrestrial origin. A belief that meteorites were ejected from terrestrial volcanoes was largely abandoned following chemical studies, although Proust (1805) proposed that volcanoes in the Antarctic would produce rocks with meteoritic properties. His idea, however received no support. (He would no doubt have been excited to learn about the recent discovery of large numbers of meteorites in Antarctica!) There were clearly different types of meteorites, some were without chondrules and were igneous rocks resembling terrestrial basalts. Maybe they came from a different lunar volcano, suggested that giant of chemisty, Berzelius. In 1834 an American astronomer, H. Olmsted, showed that the radiant of the Leonid meteor shower did not rotate with the Earth, and this put an end to any question of a terrestrial origin for meteorites (Olmsted, 1834).

Around 1840–1850 the hypothesis that meteorites came from the Moon gave way to the idea that they came from the Asteroid Belt (Humboldt, 1849). Meteorites, said Humboldt, were "the smallest of all asteroids." It had become clear that meteorites were not coming from the Moon. The lunar volcanoes were inactive and, in any case, meteorite velocities were too high. On the other hand, about a dozen asteroids were known by 1850 and it was assumed that, since asteroids were a disrupted planet, there would be many more asteroids to be detected in the future. It seemed natural to assume that a few smaller fragments could find their way to Earth. With rare exceptions (e.g. Ball, 1910), the idea that meteorites are asteroidal (or, through them, that a few might be related to their relatives, the comets) went unchallenged for 150 years. Then in the 1980s, to show that no conclusion is absolute, we discovered that a few of the igneous meteorites are from Mars and the Moon.

1.2 Museums and collectors

It was with the growth of the major industrial cities and the professionalization of science that large national collections of meteorites started to emerge. In London, Paris, Berlin, Washington, New York, Vienna, and elsewhere, collections of meteorites under the care of a professional scientist contributed to the establishment of the research field. Some of these curators, with their access to the meteorites and scientific laboratories, became leaders in the research field. It is probably the case that from about 1850 until the second half of the twentieth century the major museums were the nuclei of meteorite research, in much the same way that the NASA headquarters in Washington DC is the nucleus of the modern US space program. But in the 1960s this was about to change.

1.3 The instruments

One force that has driven meteorite studies, as with most fields of human endeavor, is the development of new instruments. Many new methods for the examination of materials "cut their teeth" on meteorite studies, beginning with the fledgling techniques of wet chemistry in 1802, largely perfected by 1834, followed by optical microscopy of geological thin sections in the 1860s to the methods of instrumental analysis of the mid-twentieth century. In the mid nineteenth century it was found that unique observations of a rock could be made by passing light through a sample (transmitted light microscopy) or by shining light on the surface of a polished sample (reflected light microscopy) (Fig. 1.3). H. C. Sorby and N. Story-Maskelyne are names associated with these developments. Sorby is well-known in the history of geology as the inventor of thin sections to be used for studying rocks and minerals and he is probably the first of the major meteorite researchers to be university-based.

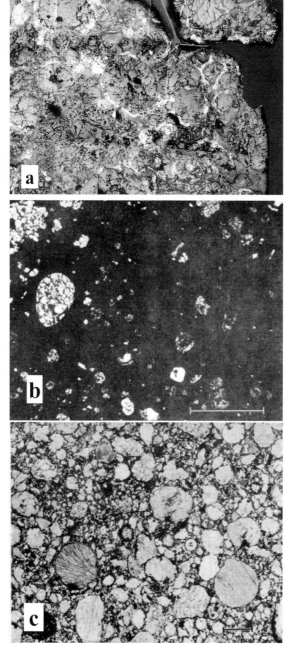

Figure 1.3 Chondritic meteorites under the microscope. (a) The Semarkona chondrite seen under the microscope with reflected light. Metal/sulfide appears white and silicates in the chondrules and matrix appear gray. The distribution of chondrules, matrix, and metal/sulfide are clearly apparent in this image. (b) The Murray CM chondrite in transmitted light. In this section, chondrules and irregular aggregates stand out clearly against the opaque matrix of hydrated silicate minerals (from Mason, 1962, p. 97). (c) The Roosevelt County H3.2 chondrite in transmitted light. Chondrules of all types are easily seen in this section (from McCoy *et al.*, 1993). In all cases the sections are about 1 cm across.

Sorby held a highly innovative position at the University of Sheffield supported by Royal Society funds so that he could be freed of the "distractions" of teaching. He used a transmitted light microscope to make the first microscopic observations of chondrules, that he described as "droplets of fiery rain from the Sun." It is significant that this was the era of astrophysics, when observations of the Sun with its sunspots, prominescences, and corona, were exciting the scientific world. Story-Maskelyne, inventor of reflected light microscopy and polished sections, was the meteorite curator at the British Museum and grandson of the famous Astronomer Royal, Nevil Maskelyne. Story-Maskelyne perfected the art of reflected light microscopy, thereby ensuring a future for metallography and opague mineral microscopy in general.

At the turn of the century spectroscopic techniques that had become popular among astronomers crept into meteorite research with the development of atomic absorption spectroscopy and through the work of the Noddacks, a husband and wife team, to systematically determine major and minor elements in meteorites. There had been hints before, but it was the American astronomer H. N. Russell, of Hertsprung–Russell Diagram fame, who convincingly showed that the chondrites contained the same elements, and in similar proportions, to those of the Sun's photosphere (Russell, 1929). A modern version of the data is shown in Fig. 1.4. In the 1950s Burbage *et al.* (1957) showed that these elements were produced by nuclear reactions associated with the evolution of stars, and now it is possible to link the isotopic properties of all meteorites with nucelosynthetic processes (Woolum, 1988).

No sooner had the idea of the similarity between chondrites and the Sun become widely accepted by researchers in the field, fine details breaking that rule began to emerge. By the 1950s enough analyses of meteorites had been acquired to be able to sift and sort through them and select only the very best, and when this was done by Nobel Laureate Harold Urey and his associate Harmon Craig in 1953, the chondrites sorted themselves into two groups – a group with high amounts of iron in its composition and with large amounts of metallic iron, and a group with low iron and low amounts of metallic iron (Urey and Craig, 1953). In this way, the H and L chondrites were born. We now recognize many such chondrite classes, all essentially solar in composition but with subtle differences in both the amount of total iron in the meteorite and in the proportion of iron in the metal state to iron in the minerals (Fig. 1.5). A few years before Urey and Craig's work, a successor to Story-Maskelyne as curator of meteorites at the British Museum, George Thurland Prior, pointed out that the less the amount of metal in chondrites the richer it was in nickel, a relationship that came to be known as Prior's Law (Prior, 1916). There appeared to be a reduction–oxidation series in the chondrites, where iron was reduced from Fe^{2+} or Fe^{3+} in the minerals to metallic Fe, or oxidized from the metallic form to the

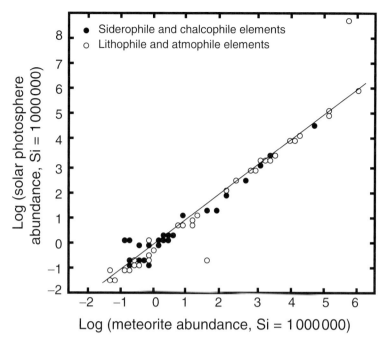

Figure 1.4 To a very good approximation, all chondrite classes are very similar to solar in composition, CI chondrites are closest, but show significant depletions in volatile elements and small depletions in siderophiles and chalcophiles. Elemental abundances in the solar photosphere are plotted against CI chondrites. Open symbols refer to lithophile (elements tending to be oxides or silicates) and atmophile elements (elements tending to be gases); closed symbols refer to siderophile (elements tending to be in the metal) and chalcophile elements (elements tending to be in the sulfides). (From Sears, 1988, who gives similar plots for other meteorite classes.)

2+ or 3+ form in the minerals. Now Urey and Craig had added that iron, probably as metal, was being removed as oxygen was being added. The removal or addition of metal is sometimes referred to as metal–silicate fractionation. In other words, we see two discrete processes; oxidation or reduction of Fe that moves samples along the diagonals in Fig. 1.5, and the removal or addition of metal which creates new diagonals (Urey and Craig, 1953; Craig, 1964). The diagonal corresponding to the Sun's Fe/Si value is shown in Fig. 1.5.

The early part of the twentieth century also saw the development of a variety of X-ray techniques such as X-ray diffraction for determining crystal structures (Young, 1926) and X-ray fluorescence for determining bulk elemental compositions (Noddack and Noddack, 1930). Eventually these X-ray techniques evolved into the electron microprobe in which a focussed beam of electrons stimulates the release of X-rays from the surface of a polished section of meteorite and these X-rays enable

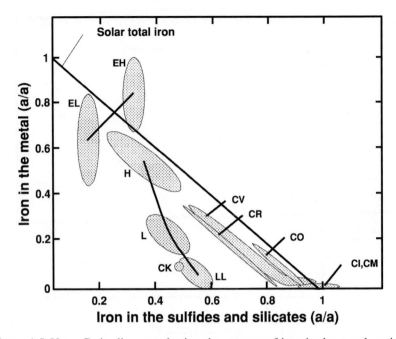

Figure 1.5 Urey–Craig diagram plotting the amount of iron in the metal against the amount of iron in the oxide and sulfide forms; in both cases the amounts are expressed as atom ratios with silicon divided by the Fe/Si atom ratio CI abundances. In this way when the ratio is one, the amount of Fe is the same in these meteorites and CI chondrites, the most primitive meteorite class. Two trends are present. When classes lie along a diagonal it means that these meteorites have uniform total iron and all that distinguishes them is the oxidation state, i.e. the amount of oxygen or sulfur that has reacted with the iron. When the meteorites lie off the diagonal there has been a loss of Fe, in other words there has been a fractionation of metal and silicates. Except for the CI and CM chondrites, the chondrite classes have experienced both processes, although to differing extents.

the identity and abundance of elements present to be determined (Castaing, 1952). Analytical techniques based on nuclear properties emerged after World War II, such as instrumental and radiochemical activation analysis (e.g. Smales et al., 1957) and isotope dilution analysis but as the twentieth century closed these gave way to mass spectroscopic techniques where the charge-to-mass ratio of ions produced from a sample are separated by magnetic and electric fields. Mass spectrometric techniques are now routinely coupled with various devices to produce an ion beam. They are also coupled with many different instruments to analyze the beam after it has passed through a mass spectrometer.

Mass spectroscopy made possible an extremely important new discipline in chondrite studies concerning their chronology, determining the time at which events occurred. The types of events that can be determined is as varied as the chemistry and physics of the isotopes available, and there are a great many. It was soon

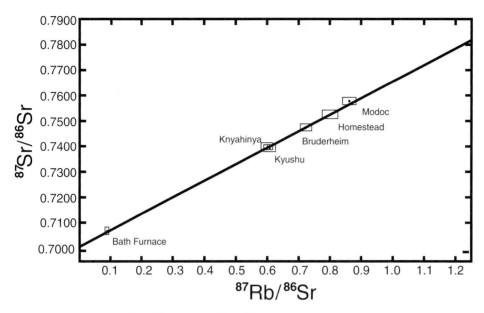

Figure 1.6 Plot of ^{87}Sr/^{86}Sr against ^{87}Rb/^{86}Sr for a suite of H chondrites indicating that they formed at the same time, 4.56 Ga ago (Kaushal and Wetherill, 1969). ^{87}Rb decays to ^{87}Sr, so the ratio of ^{87}Sr/^{86}Sr and the slope of the line increase with time. The age can be calculated from the slope on the line (slope = $e^{\lambda t} - 1$, where λ is the decay constant and t is time).

realized that meteorites are as old as radiometric estimates of the age of the Earth and astrophysical estimates of the age of the Sun. Figure 1.6 shows the results of a major study of one group of chondrites, the H chondrites, using the Rb–Sr system. Kaushal and Wetherill (1969) showed that the age of this group is 4.56 Ga (4.56 × 10^9 years). ^{87}Rb decays to ^{87}Sr with a half life of about 4.88 × 10^{10} years, so that the amount of daughter product builds up at a predictable rate and the duration of the process can be estimated. All that is needed, besides the half life, is the present and initial abundance of ^{87}Sr. The amount of Sr is a complex result of the amount initially made in the Universe and the various processes that have occurred to produce the rock, but these will also have affected ^{86}Sr – the major and stable isotope of Sr – in the same way that they affected ^{87}Sr, so we can eliminate them by taking the ratio of ^{87}Rb and ^{87}Sr to ^{86}Sr. The relationship is:

$$\left(\frac{^{87}Sr}{^{86}Sr}\right)_p = \left(\frac{^{87}Rb}{^{86}Sr}\right)_i [\exp(\lambda t) - 1] + \left(\frac{^{87}Sr}{^{86}Sr}\right)_i \tag{1}$$

where λ is the decay constant ($[\ln 2]/t_{1/2}$, where $t_{1/2}$ is half life), t is time, and the subscripts i and p mean initial and present. Thus on a plot of ^{87}Sr/^{86}Sr against ^{87}Rb/^{86}Sr such as Fig. 1.6, a group of meteorites that formed at the same time lie on a line with slope $[\exp(\lambda t) - 1]$ and with an intercept equal to the initial $^{87}\text{Sr}/^{86}\text{Sr}$ ratio. The slope on Fig. 1.6 corresponds to an age of 4.56×10^9 years. The intercept also has significance because it will have been increasing steadily since the end of element synthesis, so the later the rocks formed the higher the ratio. The intercept is another way of measuring very small differences in formation times. The Rb–Sr system can be disturbed by shock and other secondary events, and when the event is severe enough to completely reset the system the secondary event can be dated (Minster and Allégre, 1979).

Instruments have been important in driving our progress in understanding chondrites and chondrules, and a considerable body of knowledge about them now exists. However, a problem remains. In some senses, we know little more now about the origin of chondrules than we did when Sorby made his first observations (Wood, 2001). Perhaps this is in some part because the instruments at our command are not necessarily the perfect instruments for our needs. They were largely the instruments that time and circumstance enabled us to develop. The effort to learn more about the history of chondrules and chondrites, and their message for our understanding of the history of the Solar System, needs a fundamentally new way of doing business.

1.4 The space age

Probably the most important machine yet developed for further progress in understanding the origin and history of chondrules and chondrites is going to be spacecraft. For example, the Apollo program brought back nearly 400 kg of samples from six sites on the near side, equatorial region of the Moon (Heiken *et al.*, 1991). These Moon samples completely overturned many previous notions of its origin and history and of the processs occurring in the early Solar System. It is now realized, for instance, that impacts are one of the major forces in Solar System evolution. Not only does the Moon's surface show evidence for intense early bombardment by massive objects – that may also be recorded in the history of certain meteorites – but the Moon itself may be the product of a massive impact on Earth (Hartmann *et al.*, 1986).

Just as the lunar samples revolutionized our understanding of the Moon, returned samples will revolutionize our understanding of chondrites and chondrules and the early Solar System. Astronomical photometry and spectroscopy with some of the world's best telescopes has provided information on the nature of asteroids, the presumed parent bodies of chondrites, and how varied their surfaces can be. However, the asteroids remained just points of light until they were visited by spacecraft. The first images of asteroids were provided by the Galileo spacecraft on its way through the Asteroid Belt to Jupiter (Carr *et al.*, 1994). The spacecraft found a highly cratered surface, but it was not ragged and angular as predicted by many scientists. It was smooth and rounded, presumably because it was blanketed with a dusty surface layer called "regolith." A few years later the NEAR Shoemaker spacecraft flew past Mathilde and took photographs from one side. Mathilde was also remarkable for having a surface with five impact craters, each with a diameter similar to the size of the asteroid. A few years later, the NEAR Shoemaker spacecraft went into orbit around Eros, the largest of the near-Earth asteroids (Robinson et al., 2002). The NEAR Shoemaker mission was a spectacular success, with a great many structures and details on Eros being cataloged and even a small amount of compositional data being obtained from orbit and from the surface. Current technology and the space program presents us with the possibility of bringing back samples from asteroids.

Comets may be the ultimate parent body of some chondrites and certainly much of the interplanetary dust falling on Earth. The nucleus of comet Halley was viewed by a flotilla of spacecraft during its visit to the inner Solar System in 1986. The comet surface is surprisingly dark, with relatively small areas of activity at which gases and dust are being ejected into space. Features such as hills and valleys, and possibly even craters, appear to be present on the surface. The Stardust mission will fly close to the nucleus of comet Wild and will collect dust particles for return to Earth in January 2006.

The realization that chondrites have the composition of the condensable matter of the Sun was pivotol in the history of chondrite studies and is still central to our thinking. In terms of elemental composition, the similarities (and subtle differences) are well-documented if not well-understood. We also now know that the chondrite classes have many isotopic properties that distinguish them from Earth, especially in that all-important element, oxygen. The Genesis spacecraft will begin the process of comparing isotopic compositions of the Sun with those of the chondrites, including most critical information about the Sun's oxygen isotope ratios. Genesis will hover at a gravitationally stable point in space and will collect some of the stream of atoms the Sun is continually producing.

1.5 The great expeditions

Shortly after the lunar samples were returned to Earth and before spacecraft presented us with the possibility of sample return from asteroids, meteorites started to be found in huge numbers on the cold and hot deserts of Earth. The great expeditions to the Antarctic of the first few decades of the twentieth century had brought back meteorites. Their recovery is described in the journals of the first Antarctic explorers. Then, in 1969, first Japanese and then American expeditions brought back meteorites from Antarctica in unimaginable numbers. Now there is an industry of government-sponsored groups and commercial operations from Japan, the USA, and Europe returning large numbers of meteorites, not just from Antarctica but also from the prairies of North America and the deserts of North Africa and Australia. In fact, any region on Earth that is very dry – whether hot or cold – and where meteorites are well-preserved and relatively easily recovered is now being searched. As it happens, by various means these environments also cause meteorites to become concentrated in small searchable regions. Huge amounts of meteoritic material are now available to the scientist, the dealer, the collector, and the amateur through these modern expeditions to hostile regions on Earth.

Without doubt the biggest contribution of these large numbers of discoveries to meteorite research is that they enable rare and unusual meteorites to be uncovered. Large amounts of primitive asteroidal meteorites, chondrites, have been found in these collections. The first meteorite from the Moon was found in Antarctica, and now there are about twenty that are thought to come from regions of the Moon not accessible to the Apollo astronauts. About half of the meteorites originating from Mars were found in these large collections from hot and cold deserts, including Allan Hills 84001 that was supposed to contain evidence for extinct martian life.

These collections also allow us – at least in principle – to see if the nature of meteorites falling to Earth has changed since the time they fell. Some researchers argue that we should not expect to see any differences given that these meteorites fell recently in astronomical terms (within the last million years or so), but others disagree and think that they have found significant differences.

But if finding rare and relatively old meteorites is the strength of these recovery programs, one huge drawback is that these materials are all heavily weathered. They have been altered both by contamination with terrestrial materials and by the alteration of the meteorite material by terrestrial agents; water, air, and possibly microbes. It is the opinion of myself and others that the evidence for life in Allan Hills 84001 was mostly the result of weathering in Antarctica, combined with some interesting martian geochemistry. The value of the Allan Hills 84001 incident was the invigoration it gave to astrobiological studies at NASA.

1.6 Cosmic sediments

We now have 200 years of analyses of these rocks from space. We made some progress in the 1980s of placing these samples in an astronomical context, a process that was catalyzed by the discovery of tiny grains within the chondrites that are thought to predate our Solar System. We have come to a view on the history of chondrites that is described in the broadest terms by Fig. 1.7. We now believe that chondrites and their chondrules are essentially cosmic sediments. Elements created by the Big Bang were ejected into the interstellar medium to produce gas,

1.6 Cosmic sediments

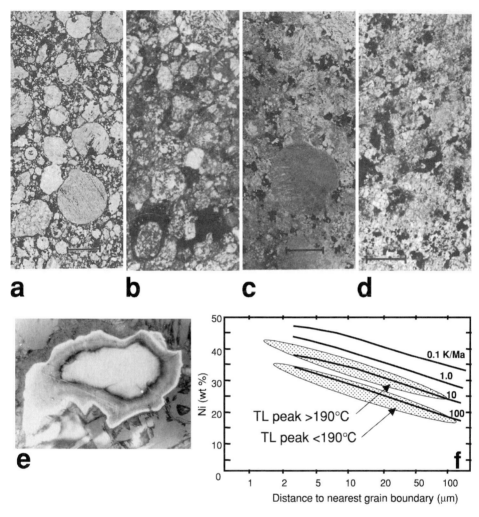

Figure 1.10 Metamorphism in chondritic meteorites. (a–d) The effect of metamorphism on the textures of ordinary chondrites as the Van Schmus and Wood (1967) petrographic type increases from (a) type 3 (Bishunpur), (b) type 4 (Cynthiana), (c) type 5 (Shelburne) to (d) type 6 (Peace River). With increasing metamorphism and petrographic type, textures become blurred and chondrules harder to delineate. (All images in the series are about 3 mm across.) (e) A metal grain (taenite, an iron–nickel alloy with face-centered-cubic structure) in the Barwell L5 chondrite showing evidence of compositional zoning that is a result of metamorphism. During metamorphism, nickel diffuses into the grain and as the temperature drops the distance the Ni can diffuse decreases, resulting in a concentration gradient. When etched with a weak acid this gradient is revealed as a characteristic pattern. (f) The amount of Ni that reaches the center of the grain is higher for smaller grains and lower cooling rates. Thus central Ni content decreases with distance from the nearest edge, the exact curve depending on cooling rate. Most chondrites cooled at 1–100 °C per million years. Data are shown (filled areas) for two categories of Antarctic meteorite, those with thermoluminescence peaks >190 °C and those with thermoluminescence peaks <190 °C (Benoit and Sears, 1992). Thermoluminescence peak temperature also depends on cooling rate.

Table 1.2 *Petrographic types for chondrite classes (Sears and Hasan, 1988).*

Property	Petrographic type							
	3.0–3.1	3.2–3.3	3.4–3.5	3.6–3.7	3.8–3.9	4	5	6
Homogeneity of olivine[a]	50		40–50	20–40	5–20	<5	Uniform	
Structural state of low-Ca pyroxene	Predominantly monoclinic →				Monoclinic >20%		ortho-rhombic <20%	
Feldspar	Absent →			Rare or absent →		<2 μm grains	<50 μm grains	>50 μm grains
Primary glass	Clear isotropic →			Turbid →	Turbid if present		Absent	
Thermoluminescence sensitivity (Dhjala=1)	<0.01	0.010–0.046	0.046–0.22	0.22–1.0	1.0–4.6	1–10	>5	
Heterogeneity of metal[b]	>17	10–17	6.0–10	2.5–6.0	<2.5		uniform	
Average Ni content of sulfide minerals	>0.5%			<0.5%				
Chondrule delineation	Very sharply defined →				well-defined		readily defined	poorly defined
Matrix texture	% transparent microcrystalline						Recrystalline matrix	
	≤20	10–20	~20–~50	>60		100		
Matrix (FeO/(FeO+MgO))[c]	>1.7	1.5–1.7	1.3–1.5	1.1–1.3	<1.1			

[a] Standard deviation of the mole % fayalite in the olivine divided by the mean fayalite expressed as a percentage.
[b] Standard deviation of the nickel content in the kamacite (wt %) divided by the mean nickel content expressed as a percentage.
[c] Divided by the same quantity for the bulk meteorite.

1.6 Cosmic sediments

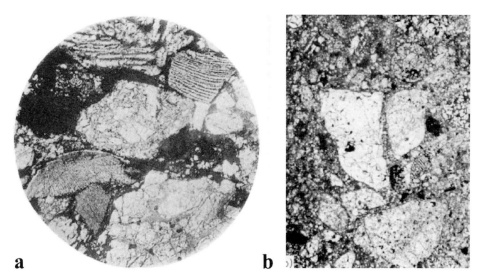

Figure 1.11 The textures of brecciated chondritic meteorites. (a) "The tuffaceous character of many chondrites is illustrated here by a photomicrograph of a portion of the Mezö Madaras" (Tschermak, 1885, Fig. 76, field of view 0.97 mm). (b) The brecciated texture of the Siena meteorite. The figure is from Kurat *et al.* (1974) who likened the texture to that of the Apollo 14 highland regolith breccias.

meteorites are rich in water while others show only the slightest traces of alteration by water-bearing fluids passing through them (Hutchison *et al.*, 1987; Zolensky and McSween, 1988; Alexander *et al.*, 1989; Grimm and McSween, 1989; Ivanov *et al.*, 1994).

Van Schmus and Wood (1967) derived the well-known criteria for assigning chondrites to petrographic types 3–6 on the basis of a number of petrographic (Fig. 1.10), mineralogical, and compositional trends. The least metamorphosed type 3 ordinary chondrites were subdivided into ten subtypes (3.0 to 3.9) on the basis of similar criteria and, more particularly by thermoluminescence sensitivity by Sears *et al.* (1980) and the scheme has been extended to CV and CO chondrites (Sears *et al.*, 1991; 1995b; see Table 1.2). Petrographic types 1 and 2 are water-rich meteorites essentially being the CI and CM chondrites in Table 1.1. Most authors regard the petrologic types as a series on increasing metamorphism, although others have argued that the chondrites were assembled from materials with differing metamorphic histories and the rocks were not metamorphosed in their present form (Fredriksson *et al.*, 1969; Fredriksson, 1983).

While much can be learned from studying the individual components under the microscope, much can also be learned by looking at the broader scale texture. Most meteorites are actually mixtures of multiple fragments, sometimes with individual fragments being of a different class to the others. In this respect the meteorites resemble volcanic materials, but we now know that it is also a characteristic of impact. Such rocks are called breccias (Wahl, 1910) and this term is sometimes qualified with additional descriptive terms. Figure 1.11 gives examples of two meteorites showing the brecciated texture. Such textures are common among rocks from the surface of the Moon (Pellas, 1973; Schultz and Signer, 1977; Rao *et al.*, 1991). Many meteorites are now thought to be impact melt rocks (Rubin 1983; 1984a; 1985). Clearly, then at some point in their history, surface processes have been important. The enstatite chondrites have a particularly interesting and unusual brecciation history (Rubin, 1984a).

In attempting to put all this together to identify the origin of chondrite classes, two processes stand out as particularly important. These are the formation of the chondrules and the metal–silicate fractionation. I will describe the many varied ideas that have been presented to explain these processes and I will present admittedly controversial arguments that both processes occurred on the surfaces of their parent bodies.

2
Potential meteorite parent bodies

2.1 Asteroids as potential meteorite parent bodies

2.1.1 The difficulties

Most meteorite researchers are confident that the near-Earth asteroids are the immediate parent bodies of the chondritic meteorites. Asteroid astronomers generally accept that most of the near-Earth asteroids originate in the Main Asteroid Belt, although some, probably around 10% (estimates have ranged as high as 50% in the past) are inert residues of comet nuclei (Wetherill and Chapman, 1988). However, when it comes to more precise links between asteroids and meteorites there are problems. As discussed below, the spectra do not fit; the commonest kinds of meteorites have spectra that match very few asteroids, while the majority of asteroids have spectra that match very few meteorites. Either the asteroids are camoflaging themselves or most of the meteorites are coming from very few sources. Just to compound the mystery, Eros, which has many of the compositional properties of an LL chondrite, has significant differences. We shall return to this below.

Before getting into details, it is important to stress the difficulties of linking asteroids, based on astronomical and remote sensing data, with the meteorites we analyze in the laboratory (Bell *et al.*, 1989; Sears, 1998; Meibom and Clark, 1999).

First, we are only looking at the very surface of the asteroids with our photometric and spectroscopic techniques. It is possible that the interior is quite different from the surface where space conditions may have caused alterations. Certainly, when we use geophysical methods to determine asteroid densities they are very much less than those of meteorites, suggesting considerable porosity or the presence of water. Asteroid densities are 1.2–2.8 g/cm^3 for chondritic-like asteroids (Table 2.1), the mean density for S asteroids is $\sim$2.4 g/cm^3 and for C asteroids $\sim$1.8 g/cm^3. In comparison, meteorite densities are $\sim$3.4 g/cm^3 for ordinary chondrites and $\sim$2.0 g/cm^3 for C chondrites (Table 2.2). Either all asteroids are essentially C chondrites whatever their surface chemistry, or S asteroids have considerable water and

Table 2.1 *Asteroid densities (Britt et al., 2002).*[a]

Asteroid	Class	Density (g/cm^3)
1 Ceres	G	2.12±0.04
2 Pallas	B	2.71±0.11
4 Vesta	V	3.44±0.12
10 Hygeia	C	2.76±1.2
11 Parthenope	S	2.72±0.12
15 Eunomia	S	0.96±0.3
16 Psyche	M	2.0±0.6
20 Massalia	S	3.26±0.6
22 Kalliope	M	2.5±0.3
45 Eugenia	F	$1.2^{+0.6}_{-0.2}$
87 Sylvia	P	1.62±0.3
90 Antiope	C	1.3
121 Hermione	C	1.96±0.34
243 Ida	S	2.6±0.5
253 Mathilde	C	1.3±0.2
433 Eros	S	2.67±0.03
704 Interamnia	F	4.4±2.1
762 Pulcova	PC	1.8±0.8
804 Hispania	PC	4.9±3.9
1999 KW4		2.39±0.9
2000 DP107	C	$1.62^{+1.2}_{-0.9}$
2000 UG11		$1.47^{+0.6}_{-1.3}$

[a] Mean densities (in g/cm^3): C ~1.8; S ~2.4; V ~3.4.

porosity. Water can be detected on asteroids from the strength of the 3 μm absorption feature (Table 2.3). As expected, water has been detected on one-third to one-half of the C and related B asteroids which are expected to be CI-like in composition, and not in the S asteroids whose surfaces are dry.

The second difficulty in comparing the terrestrial meteorite flux with asteroid class distribution is that there is no guarantee that the meteorites reaching our museums are typical of what is in the Asteroid Belt, and every indication that it is not. What is sent to Earth from the Asteroid Belt depends on what gets impacted and sent on an Earth-crossing trajectory, and this depends to some extent on original location. In addition, the atmosphere destroys all but the toughest material. We have ample evidence in the form of theoretical but well-understood calculations, observed falls of fragile meteorites – where huge meteorites enter the atmosphere but little is recovered (Revelstoke and Taggish Lake are examples of this) – and military data for "air blasts" that suggest very large numbers of large objects are being destroyed high in the atmosphere. In short, there is probably no reason to assume that what lands on Earth in the form of meteorites is representative of the

Table 2.2 *CI and CM chondrites and their water contents and densities.*

		Water		Porosity		Bulk density	
Meteorite	Class	wt %	ref.[a]	vol. %	ref.[a]	g/cm^3	ref.[a]
Alais	CI	19.62	1	2	3	–	–
Ivuna	CI	43.47	1	–	–	–	–
Orgueil	CI	16.9	2	11.3	3	2.11±0.12	3
ALH 81302	CM	12.94	2	–	–	–	–
ALH 83100	CM	13.38	2	–	–	–	–
Benten	CM	10.7	2	–	–	–	–
Cold Bokkeveld	CM	15.3	1	12.9	3	2.31	3
Erakot	CM	19.26	1	–	–	–	–
Essebi	CM	9.9	1	–	–	–	–
Haripura	CM	36	1	–	–	–	–
Mighei	CM	2.16	1	28.2	3	1.94±0.03	3
Murchison	CM	10.09	2	17.1	3	2.37±0.02	3
Murray	CM	12.51	1	–	–	–	–
Nawapali	CM	16.56	1	–	–	–	–
Santa Cruz	CM	10.44	1	30.3	3	1.79	3
Staroje Boriskino	CM	0.99	1	–	–	–	–
Yamato 791824	CM	12.5	2	–	–	–	–
Yamato 793321	CM	9.23	2	–	–	–	–

[a] References: 1, Wiik (1969); 2, Jarosewich (1990); 3, Britt and Consolmagno (2002).

Table 2.3 *Asteroids with spectroscopic evidence for water on their surface (Vilas, 1994).*

Asteroid class	Number in sample	Number with water (%)	Mean albedo of hydrated asteroids
A	5	0	–
B	18	33.0	0.049
C	128	47.7	0.051
D	35	0	0.041
E	12	0	–
F	36	16.7	0.042
G	7	85.7	0.077
I	3	0	–
M	31	6.5	0.155
P	36	8.3	0.036
Q	1	0	–
R	1	0	–
S	201	0.5	0.095
T	6	0	–
V	1	0	–
X	54	5.6	–

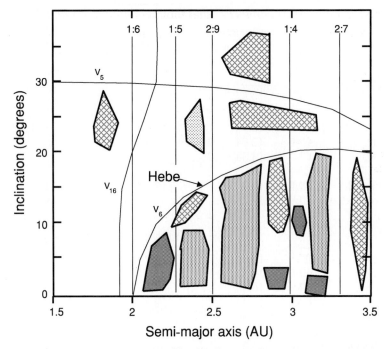

Figure 2.1 The orbital inclination of the asteroid orbits against their semi-major axis with the resonances indicated. The intensity of the filling for the different fields reflects the density of asteroids plotting in that region (dark vertical hatching > light vertical hatching > cross-hatching > stippled). Certain locations on this plot seem to be forbidden. The vertical lines are the famous Kirkwood Gaps, where the period of the orbit around the Sun for the asteroids is in simple ratio to Jupiter's 12-year period, for example asteroids at 2 AU have a solar orbit taking 2 years, in a 1:6 ratio with Jupiter. These are the "resonances" where long-term interactions by Jupiter sweep the region clear of asteroids. Similar but more complex gravitational interactions occur with Saturn, leading to the resonances labeled v that appear as curves on this plot. The resonances act as "escape hatches" through which asteroids can escape the Main Belt and become Earth-crossing. The S asteroid Hebe plots near two of these resonances and may be the source of the H chondrites.

materials in the Asteroid Belt, although a considerable amount of effort has been spent in reconciling the observations.

2.1.2 Orbital mechanics

The variety of orbits occupied by asteroids is best summarized on a plot of two of their orbital elements, the inclination of the plane of their orbit to that of the Earth's orbit and the semi-major axis of their orbits (Fig. 2.1).

On such a plot, a number of fields and tendencies are observed (Binzel *et al.*, 1989). Asteroids tend to avoid certain regions, where gravitational interaction with

Jupiter and Saturn sweep them away. These are termed the orbital resonances and they are shown in Fig. 2.1 either as vertical lines, the well-known Kirkwood Gaps (when only Jupiter is involved), or they are curves (when Saturn is involved).

The subject of the orbital evolution of objects in the Asteroid Belt and how meteorites might have been transfered from the asteroids to Earth has been reviewed many times (Davis *et al.*, 1979; Greenberg and Chapman, 1983; Wetherill, 1985; Wetherill and Chapman, 1988; Greenberg and Nolan, 1989). Until the mid-1980s it seemed difficult to understand how the Asteroid Belt could be providing the observed meteorite flux to Earth, but with the discovery of chaotic zones around resonances (Wisdom, 1985) this difficulty has been overcome and the resonances are now regarded as "escape hatches." There have been a great many studies of how orbits evolve after ejection of asteroid fragments into a resonance by, for example, impact. The sixth largest asteroid Hebe, a proposed parent body for the H chondrites (Gaffey and Gilbert, 1998) is located near both the 2:9 resonance with Jupiter and the v_6 resonance with Saturn. It seems possible that the parent body of the L chondrites – where there are signs of an extremely violent event throughout most of the class – no longer exists, being completely destroyed by the impact that sent pieces to Earth.

In the last decade or so, there has been considerable interest in another process for moving objects from the Asteroid Belt to the vicinity of Earth, namely the Yarkovsky effect (Farinella *et al.*, 1998). Instead of being a stochastic process capable of sending very large numbers of large fragments from a few events, the Yarkovsky effect moves all objects smaller than ~ 1 m at the rate of about 0.001 AU per million years, some objects are moved inwards and some are moved outwards. The effect results from the Sun-facing side of the asteroid being warmer than the opposite side and radiating more energy. Because of the rotation and thermal lag, the direction of rotation will determine whether movement is inwards or outwards. This process is capable of explaining the background of small, highly diverse objects reaching Earth, but not the major classes.

2.1.3 Spectroscopic classes of asteroids

Some sixteen classes of asteroids are now recognized by asteroid astronomers based on the spectra of reflected sunlight. The biggest class has flat spectra and the objects also have low albedo; in both these respects they resemble carbonaceous chondrites (Fig. 2.2). They are referred to as the C class asteroids and there are a number of similar classes. Asteroids with absorption bands due to the silicates olivine and pyroxene and high albedos are termed the S asteroids (Larson and Veeder, 1979; Wetherill and Chapman, 1988; Gaffey *et al.*, 1993a; 1993b). C asteroids tend to predominate in the outer regions of the Main Asteroid Belt, while S asteroids

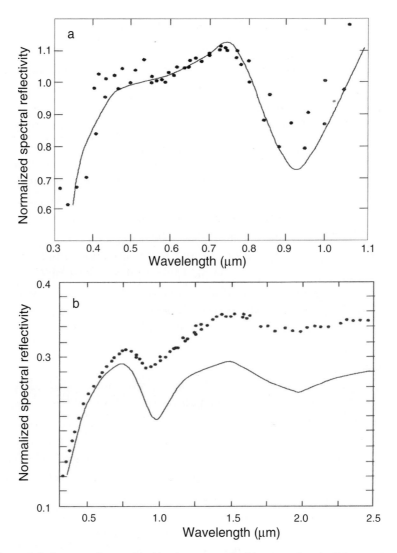

Figure 2.2 Spectra of asteroids (dots) compared with meteorites (solid curves). (a) The spectra of Vesta and a howardite meteorite compared (McCord et al., 1970). (b) The spectra of an S asteroid and an ordinary chondrite compared (Chapman, 1976). Vesta is one of the largest asteroids and has a bulk density much higher than most asteroids. It is thought to be differentiated with its surface covered in basalts. This would be consistent with eucrites (and the related howardites and diogenites) coming from Vesta. These non-chondritic meteorites are not discussed in this book, but are shown here to demonstrate the potential success in matching asteroid spectra. Space weathering on Vesta should resemble space weathering on the Moon, being similar target materials, yet the meteorite match is good and there is no sign that this has altered the spectrum. The S spectra is of high intensity and shows absorption features similar to the meteorite but they are not identical, the asteroid spectra being more intense at longer wavelengths (i.e. "redder"). The question is whether the differences preclude the S asteroids being the parent objects of ordinary chondrites or whether the conditions of space or some other process has altered the asteroid spectrum. Not shown are the spectra of a C asteroid which is of low intensity and featureless, but consistent with C chondrites coming from C asteroids.

Table 2.4 *Asteroid–meteorite matches (Gaffey et al., 1993a).*

Type	Major mineral phases[a]	Possible meteorite analogy[b]
V	Pyroxene +/−Feldspar	Eucrites, howardites, diogenites
A	Olivine +/−FeNi metal	Oilvine achondrites pallasites, *olivine–metal partial melt residues*
E	Enstatite (<Fsl)	Enstatite achondrites (aubrites), iron-bearing enstatite (Fs2–4), *Fe-bearing aubrites (Fs2–4)*
R	Olivine + orthopyroxene	*Olivine–pyroxene cumulates, olivine–pyroxene partial melt residues*
M	Metal +/−enstatite hydrated silicates + organics?	Iron meteorites, enstatite chondrites
S		
S(I)	Olivine ⋙ pyroxene (+/−FeNi metal)[d]	Pallasites, pyroxene-poor ureilites, pyroxene-poor brachinites, *olivine–metal partial melt residues*,
S(II)	Olivine ≫ clinopyroxene (+/−FeNi metal)[c], (0.05 < cpx/(ol+cpx) < 0.20)[d]	Cpx-bearing ureilites, Cpx-bearing brachinites, *olivine–Cpx cumulate, Cpx-bearing pallasites, highly metamorphosed, C-type assemblages*
S(III)	Olivine > clinopyroxene + orthopyroxene (+/−FeNi metal)[d]	Cpx- and opx-bearing ureilites
S(IV)	Olivine + orthopyroxene (+/−FeNi metal)[d], (0.20 < opx/(ol+opx) < 0.50)	Opx-bearing ureilites, lodranites, winonites & IAB iron, H, L, LL chondrites
S(V)	Olivine – clinopyroxene (+/−FeNi metal)[d]	Lodranites, *Cpx-basalt intrusions into H-chondrite matrix*
S(VI)	Olivine – orthopyroxene (+/−FeNi metal)[d]	Siderophyres (Steinbach), lodranites, winonltes & IAB irons, *subsolidus-reduced chondrites, anorthosites*
S(VII)	Pyroxene > olivine (+/−FeNi metal) (orthopyroxene > clinopyroxene)	Mesosiderites siderophyres (Steinbach), lodranites, winonites & IAB irons, *Cpx-poor mesosiderites, subsolidus-reduced chondrites, anorthosites*
Q	Olivine + pyroxene (+ metal)	Ordinary chondrites
C	Iron-bearing hydrated silicates	CI1 & CM2 chondrites, *dehydrated CI1 & CM2 assemblages*
B	Iron-poor hydrated silicates	*Partially dehydrated highly leached CI1-type assemblages*
G	Iron-poor hydrated silicates	*Highly leached CI1-type assemblages*
F	Hydrated silicates + *organics*	Organic-rich CI1 & CM2 assemblages
P	*Anhydrous silicates + organics*	Olivine–organic cosmic dust particles
D	*Organics + anhydrous silicates*	Organic–olivine cosmic dust particles
T	*Troilite (FeS) (+ FeNi metal)*	Troilite-rich iron meteorites
K	Olivine + opaques	CV3/CO3 chondrites
Z	*Organics (+ anhydrous silicates)*	Organic-rich cosmic dust particles

[a] Mineral species or assemblages in italic font are inferred from spectral properties which are not specifically diagnostic.
[b] Analogs in italic font have not been found or presently identified in meteorite collections.
[c] Characterizations from Gaffey *et al.* (1993b).
[d] Metal abundance is poorly constrained and appears to be highly variable.

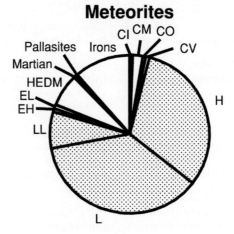
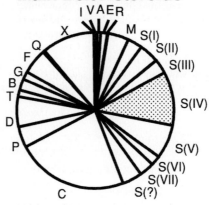
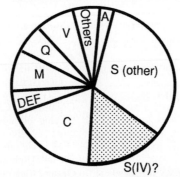

Figure 2.3 Pie charts showing the distribution of asteroids and meteorites over the classes showing the "S asteroids paradox" (Sears, 1998) – that while ordinary chondrites are abundant falls to Earth asteroids matching their spectra are rare.

2.1 Asteroids as potential meteorite parent bodies

predominate in the inner belt. The S class actually shows considerable variability in its spectral properties and has been further subdivided into subclasses, bringing the total number of asteroid spectral classes to 21. Gaffey *et al.* (1993b) have summarized progress in identifying asteroid mineralogies from their reflected light spectroscopy and tried to link them with the meteorites. There is much uncertainty over attributions of broad and sometimes weak bands in these solid state spectra, and sometimes it is necessary to invoke materials that are not found in meteorites, but Table 2.4 represents the most optimistic and thorough attempt to describe the mineralogy of asteroids and their links with meteorites. The classic best fit is Vesta and the eucritic meteorites (McCord *et al.*, 1970), and nobody doubts this association (Fig. 2.2a). Gaffey and Gilbert (1998) have suggested that 6 Hebe is the parent body for the H chondrites, but its spectrum differs from H chondrites in a manner similar to the ordinary chondrite example in Fig. 2.2b. A component is required that will increase the reflectivity at long wavelengths ("redden" the spectra). It has been suggested that metal has been added to the surface and since the IIE iron meteorites are thought to be related to the H chondrites, Gaffey and Gilbert (1998) suggested that IIE was present on the surface of the Hebe H chondrite parent asteroid.

The distribution of meteorites and asteroids over their various classes are shown as pie charts in Fig. 2.3. Less than 1% of asteroids have spectra readily attributed to ordinary chondrite material, yet 95% of the meteorites falling on Earth are chondrites. On the other hand, while C asteroids are almost as common as S asteroids, only a few percent of meteorites falling on Earth are C chondrites. Of course, the explanation for this difference might lie in the non-representative nature of the materials reaching our museums, as discussed above, but considerable effort has been spent looking for other explanations.

Since spectra only refer to the surface of the asteroid, we might consider processes that alter the surface. It is well-known that certain meteorites are surface material, they contain solar wind − implanted gases and charged particle tracks that would be impossible if the meteorites had been buried only a few centimeters deep. Furthermore they have a very characteristic texture of normal light clasts of

---

← On the other hand, C chondrites are rare falls on Earth, but asteroids with spectra matching those expected for C chondrites are common. Even if one is very liberal in the comparison, probably only the S(IV) can be said to match ordinary chondrites. Possible explanations are that the surfaces of asteroids are being altered by the space environment or that the flux of meteorites to the surface of the Earth is not representative: perhaps ejection to Earth requires rare single events like appropriate impacts so the terrestrial flux is random, or passage through the atmosphere destroys all but the strongest meteorites. The similarity of the near-Earth asteroid and the Main Belt distributions suggests it is an atmospheric effect.

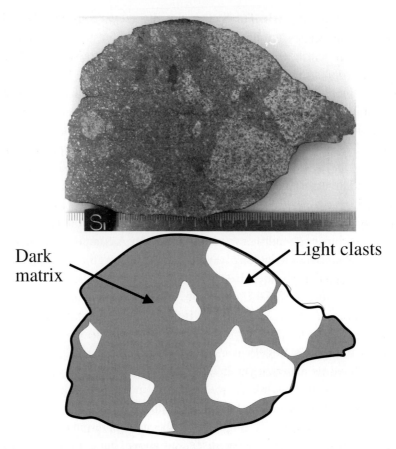

Figure 2.4 The Fayetteville gas-rich regolith breccia. The light clasts have all the properties of normal H chondrites, but the dark matrix is comminuted clast material to which has been added fragments of other classes (usually CM chondrites), shock melted fragments, and considerable amounts of trapped solar wind. Crystals in the dark matrix also have high densities of charged particle tracks caused by solar particles entering the minerals at high energy. Such gas-rich regolith breccias were clearly located on the surfaces of their parent bodies as dust and debris prior to becoming the coherent rocks we see today. These types of meteorite are present in all the chondrite classes, although not in great numbers.

centimeter-sized fragments of H-chondritic material enclosed in a dark matrix of crushed clast material rich in trapped solar wind gases and charged particle tracks (Fig. 2.4). Bell and Keil (1988) and Britt and Pieters (1991) have suggested that the gas-rich regolith breccias are likely to be representative of the surface materials on meteorite parent bodies and may partly explain the meteorite–asteroid mismatch.

The surface of the Moon has been studied intensely as a result of the Apollo program and the processes occurring there have been very thoroughly documented (Heiken *et al.*, 1991). The process is generally referred to as "space weathering" and involves the production of glassy, dusty fragments called agglutinates and the formation of submicroscopic iron particles by the evaporation, reduction, and condensation of iron-bearing minerals (McKay *et al.*, 1991). Thus the iron-normalized magnetic susceptibility of the lunar regolith correlates with parameters of surface exposure such as solar wind gases, carbon, radiation tracks, and thermoluminescence sensitivity (Hörz *et al.*, 1991; Batchelor *et al.*, 1997). There has been considerable controversy over the role of radiation damage in this process and the relative importance of chemical and physical processes. Some of the effects shown on the surface of lunar grains and attributed to recondensation could be due to radiation damage (Borg *et al.*, 1980).

The lunar regolith breccias and soils show many of the same trends as the gas-rich regolith breccias with the dark matrix–light clast structure and some authors are convinced that a lunar-style space weathering is occurring on asteroids (Pieters *et al.*, 2000). It might be this process that is masking the number of asteroids with ordinary chondrite compositions. Binzel *et al.* (1996) found a gradation in spectra from S type to ordinary chondrite type among small asteroids and argued that this is evidence for a gradational process like space weathering, and Chapman (1996) has found differences in spectra of various regions on Ida that can be attributed to surface age and therefore degree of space weathering.

However, there are problems. Meteoritic regolith breccias do not contain agglutinates and Hörz *et al.* (1991) and Hörz and Cintala (1997) have stressed the differences in the conditions on the surfaces of asteroids and the Moon. These mainly concern location and parent body size. Keplerian velocities are much higher in the Asteroid Belt than in the vicinity of the Moon so one might expect higher impact velocities in the belt, especially since orbital interactions with Jupiter scatter the orbits. Furthermore, the gravitational field of an asteroid is negligible compared to that of the Moon and this would affect the nature of surface regolith processing on asteroids. For example, one might expect greater escape of material from an asteroid, much more material to be excavated during a typical impact, and much larger flight times for ejecta that returns to the surface. This might explain the lack of agglutinates on brecciated meteorites because the longer flight times would mean that the glass droplets would solidify before they have time to splash onto the surface to produce aggluntinates. Another major difference between asteroid surfaces and the lunar surface, of course, concerns the composition of the target rock. The Moon's surface is very dry and refractory while the asteroids can be volatile-rich, perhaps even containing water, and are not refractory in composition. It is not clear what these compositional differences would do to the space weathering process. So, in addition to expecting differences in surface weathering between the asteroids and the Moon, one might expect very big differences in the nature of space weathering on different asteroid types. Hörz and Cintala (1997) remark on how little is known about asteroid space weathering processes compared with lunar processes. This difference is surely a result of having samples from lunar regolith and not from the asteroid regolith.

2.1.4 Evidence for meteorite parent bodies

If the meteorites come to Earth as fairly small individuals, capable of containing cosmic-ray-produced nuclides throughout their mass, how do we know that they were not always small objects? How do we know that they came from larger "meteorite parent bodies" that we equate with the asteroids? From the perspective of the meteorites, the best evidence for them having come from larger parent bodies is provided by the age distributions of chondrites, specifically the cosmic ray exposure ages (Grabb and Schultz, 1981; Marti and Graf, 1992; Fig. 2.5) and Ar–Ar ages (Bogard, 1994; Fig. 2.6).

A particularly large number of H chondrites, perhaps all, have cosmic ray exposure ages of ~8 Ma. Since the size of a meteorite must be of the order of one meter for cosmogenic reactions to become important, many (perhaps most) of the H chondrite meteorites must have been part of a single parent object until fragmentation occurred about eight million years ago. They were effectively "shielded" from cosmic rays at that time. The LL chondrites also show a peak in the cosmic ray histogram, at 17 Ma, so a similar argument can be made for them. The L chondrites do not show a single peak, but there have been suggestions that a number of poorly defined peaks might be present. It is important to note that all factors affecting cosmic ray exposure ages (gas loss, shielding effects, laboratory errors) serve to diminish the peak, so the number of samples in the peak is a lower limit to the number of meteorite samples involved in the event.

While the L chondrites do not produce distributions with unequivocal peaks, the Ar–Ar ages for meteorites from this class show a strong peak at ~500 million years (~0.5 Ga) indicating a major degassing event at that time (Heymann, 1967; Turner, 1988; Bogard, 1994). Again this event was shared by a particularly large fraction of the members of this class, and it is possible that they were all involved. Some insight into the nature of the event that reset the Ar–Ar clock at ~500 million years is that most of the meteorites plotting in the peak show signs of a violent event. They are often shock-blackened, many contain "veins" – cracks filled with metal and sulfide that appear to have been forced through them as a melt, olivine grains with strange optical properties (called undulatory extinction), and other effects (Stöffler et al., 1988). It seems that the parent body of the L chondrites underwent a violent shock event ~500 million years ago (such as a collision between two asteroids) and this caused pressure and temperature transformations in the meteorites. Of course

2.1 Asteroids as potential meteorite parent bodies

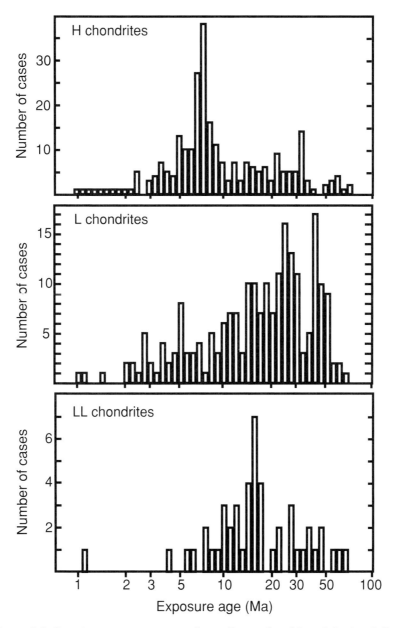

Figure 2.5 Cosmic ray exposure ages for ordinary chondrites (Marti and Graf, 1992). Cosmic ray interactions in the meteorite produce certain nuclides whose abundance can be used to estimate the cosmic ray exposure age. There is uncertainty as to whether the L chondrites show any significant peaks, but the H chondrites show a very sharp peak at 8 million years and the LL chondrites show a less sharp peak at 17 million years. In order to have become exposed to cosmic rays, meteorite fragments must be less than a meter or so in size. Thus until 8 million years ago (17 million years for the LL chondrites), many or perhaps all of the H (or LL) chondrite fragments must have been part of a common parent body.

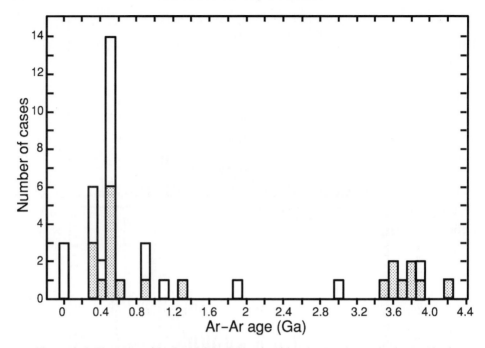

Figure 2.6 The ages of L chondrites determined by the Ar–Ar method, a method of precisely obtaining K–Ar ages (Bogard, 1994). The peak at 0.5 Ga indicates that until this point in time a large proportion of the meteorites, probably all of them, were part of a single common parent object and that 0.5 Ga ago a violent heating event caused complete loss of Ar that reset the K–Ar clock. The event was so violent that it also caused the meteorites to become shock-blackened and minerals to become melted and otherwise altered. Even the peak at ~3.8 billion years for unshocked and unheated meteorites presumably reflects some loss of Ar. Dating techniques that do not rely on a gas that can diffuse out of the sample give formation ages for these meteorites of ~4.6 billion years.

one of the major temperature effects is to drive off the Ar and other gases in the meteorites and thus reset their K–Ar chronometers.

2.1.5 Numbers of meteorite parent bodies

The existence of a number of classes of chondrites with remarkably isochemical properties is sometimes regarded as an indication that there were discrete meteorite parent bodies. Thus some authors have suggested that there are over 100–150 parent asteroids represented in our meteorite collections (Burbine et al., 2003). However, until we can understand the processes giving rise to these groupings, or the typical heterogeneity that can be found on a single asteroid, these ideas should be considered no more than reasonable suggestions.

2.1.6 Accretional history of asteroids

Petrographic evidence concerning the nature of the accretion process can also be gleaned from the meteorites. Some authors have argued that "accretion" and metamorphism were contemporaneous, but it is important to distinguish between the primary accretion of dust grains in the nebula to produce meter- and kilometer-sized objects and the reworking of an existing surface; this is not a trivial task. It can probably be concluded fairly safely that at least some chondrules were pliable when they were finally assembled into the meteorites we now observe (Hutchison et al., 1979; Hutchison and Bevan, 1983).

Laboratory simulation experiments have proved extremely valuable in understanding possible surface processes on comets (Sears et al., 1999b) and the Moon (Hörz and Cintala, 1997), yet few such experiments have been performed in an effort to understand asteroids and chondrites.

2.1.7 Thermal models and the history and structure of meteorite parent bodies

Thermal models have the potential to provide information on the size and structure of parent bodies and the location of the meteorites in those bodies. Goldstein and Short (1967) and Wood (1964; 1967a) pioneered the estimation of cooling rates from metal structures and concluded that the iron meteorites and chondrites were coming from parent objects several hundred kilometers in size. The major uncertainties in these results concerned the geometry of the kamacite and the thermal properties of parent body, and assumptions that the kamacite grew as spindles rather than infinite plates decreased cooling rate estimates by three orders of magnitude.

Similarly, better estimates for the thermal properties of chondritic material (e.g. Matsui and Osaka, 1979) and peak metamorphic temperatures (McSween et al., 1988), allowing for regoliths and megaregoliths and for the long-range porosity of the whole parent body, have resulted in a series of thermal models in which chondrites have moved ever closer to the surface of that body (Miyamoto et al., 1981; 1986; Haack et al., 1990; Bennett and McSween, 1996) until Akridge et al. (1998) found that chondrite cooling rates and peak temperatures are consistent with types 3–5 being from the regolith and type 6 from the megaregolith of their parent bodies (Fig. 2.7a).

King et al. (1972b) suggested a reasonable thermal history of impact-emplaced, hot, base surge, and fall-back deposits and a possible relationship with the petrologic types of Van Schmus and Wood (1967) (Fig. 2.8). Depth of burial determines peak temperatures and cooling rates, and thus degree of welding and crystallization, heterogeneity of mineral compositions, distinctness of chondrules, and contents of volatile elements. While King et al. (1972b) had fall deposits of an impact,

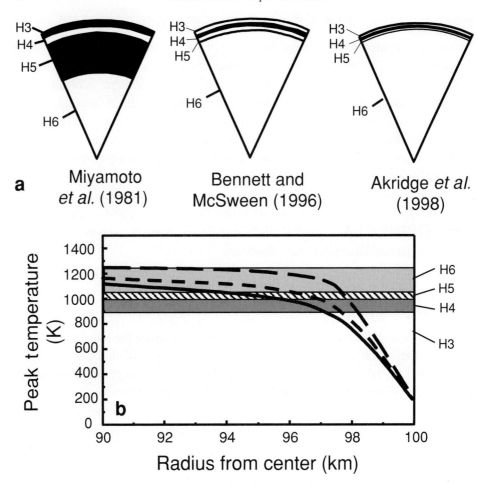

Figure 2.7 (a) Thermal models assuming spherical symmetry in the degree of metamorphic alteration and thus petrographic type. This is sometimes referred to as an "onion skin" structure. The Miyamoto et al. (1981) calculations suggest that type 6 chondrites were buried deep in the parent body, but revisions in thermal conductivity and estimates of metamorphic temperatures brought all types closer to the surface in the Bennett and McSween (1996) calculations. Akridge et al. (1998) included the effect of a regolith and a megaregolith, and results suggest that all petrographic types could have formed in the regolith or megaregolith of the parent body. (b) Temperature profiles from the work of Akridge et al. (1998) with the petrologic types superimposed for the outer 10 km of a 100 km (Hebe-sized) parent object.

such a qualitative description could apply to any regolith-like layer many tens, or even hundreds of meters thick, such as the megaregolith of Akridge et al. (1998) (Fig. 2.7b). Miyamoto et al. (1986) show that the compositional profiles in olivine in chondrules is consistent with burial in a hot ejecta blacket several tens of meters thick.

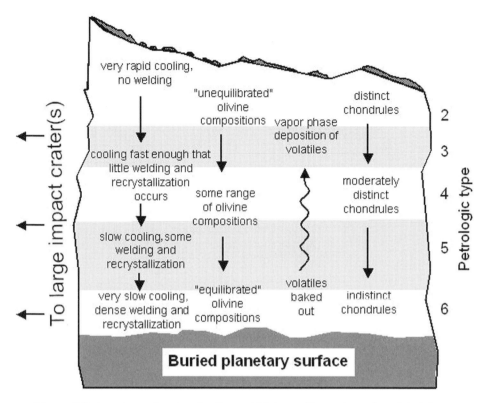

Figure 2.8 A cartoon showing the thermal history of impact-emplaced, hot, base surge, and fall-back deposits and a possible relationship with the petrologic types of Van Schmus and Wood (1967) that describe metamorphic alteration in the chondrites. Depth of burial determines peak temperatures and cooling rates, and thus degree of welding and crystallization, heterogeneity of mineral compositions, distinctness of chondrules and contents of volatile elements (after King et al., 1972b).

But what about the structure of the asteroid as a whole? Traditionally we have thought in terms of two possibilities, the "onion skin" structure (Fig. 2.9a) and a rubble pile structure (Fig. 2.9c). The onion skin model assumes that the heat source is internal and petrographic type decreases towards the outside in a global, spherically symmetrical manner. Distances, temperatures, and cooling rates are given by calculations like those described in Fig. 2.7a.

Some authors appear to have found a weak correlation between cooling rate and petrographic type, suggesting that the more metamorphosed material was buried more deeply, consistent with the onion skin structure. But data are meager and correlations are weak. Other workers have argued that the lack of coherent metal composition and the brecciated textures of meteorites suggest that the metal grains did not cool in their present location in the meteorite and that the meteorite must

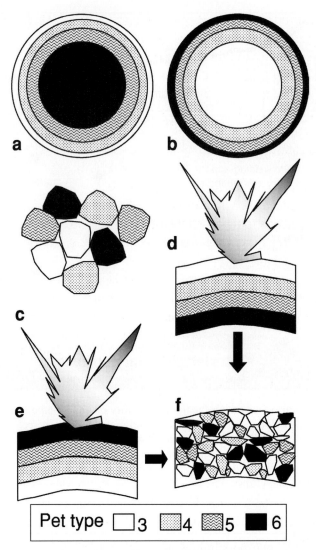

Figure 2.9 Cartoons describing possible parent body structures for chondrites. (a) The "onion skin" structure (concentric layers) in which the heat source is internal and petrographic type decreases towards the outside in a global, spherically symmetrical manner. (b) An onion skin structure in which heating is external, for example by impact, and petrographic type would decrease with depth, although perhaps in a fairly irregular fashion since impact heating need not be uniform. (c) A "rubble pile" idea in which chunks of varying petrographic type are randomly assembled. In this case, the chunks could be the result of fragmentation of older onion skin bodies or the chunks could have been heated to differing degrees by varying amounts of ^{26}Al. Internally heated onion skin structures (d) or externally heated onion skin structures (e) could also produce rubble pile like structures (f) as a result of impact brecciation.

have been a rubble pile (Scott and Rajan, 1981; Scott et al., 1985; Taylor et al., 1987) However, there are many other reasons why metal grains would not show coherent metal composition plots and the small size of the meteorites in comparison with their asteroidal parents, and the ubiquitous nature of brecciation on a wide variety of distance scales should be borne in mind. The rubble pile proposal was prompted by the realization from energy calculations that some asteroids must have been totally disrupted by impact and subsequently reassembled (Farinella et al., 1982). Many asteroids studied by radar techniques appear to consist of two gravitationally bound bodies covered by a continuous layer of dust. It was suggested that these objects were an asteroid that had become completely fragmented by impact but which had reassembled itself. The discovery of evidence for ^{26}Al in the meteorites provides an explanation for different thermal histories for fragments that are in the same region of space, since those that formed early would have more ^{26}Al heating than those that formed later.

In principle it should be possible to distinguish between the onion skin and rubble pile structures by precise dating. For the rubble pile, petrographic type 6 accreted early, then type 5, then type 4, and finally type 3. For the onion skin structure, the reverse is true – type 3 should form first, then 4, then 5, and finally 6 since materials at the lower temperatures cease to undergo physical and chemical changes first as the body cools from the outside inwards. These measurements require that dating techniques are applied very precisely, but at least two research groups (in Paris and Heidelberg) have found ages that correlate with petrographic types and therefore favor the onion skin structure, at least for the H chondrites (Trieloff et al., 2003).

There are additional possibilities for meteorite parent body structures. One is that the body was externally heated by impact or solar effects (Fig. 2.9b). This idea was rejected twenty years ago but needs to be revisited in the light of new discoveries concerning the importance of regoliths. Another possibility is that thermally zoned outer layers were later mixed by impact brecciation (Figs. 2.9c and 2.9d) to produce the non-coherent composition–dimension plots (Fig. 2.9f).

Crucial to these arguments about parent body structure is the heat source. Internal heating by radioactive nuclides would produce concentrically zoned parent bodies but with cosmic abundances of U, Th, and K. The parent bodies have to be larger than 800 km and this seems very unlikely in view of the sizes of asteroids and current views on their evolution. The discovery of ^{26}Mg excesses that correlate with ^{27}Al provides evidence that ^{26}Al was present when the meteorites formed, and if universally present throughout the meteorite parent bodies would have melted any object greater than 5 km in size (Lee et al., 1976; Herndon and Herndon, 1977). However, this assumption of homogeneous distribution of ^{26}Al can be questioned. Alternate heat sources, such as impact heating (Kaula, 1979) or heating by electrical

Table 2.5 *Asteroids[a] visited by robotic spacecraft and sources of information on them.*

Asteroid	Spacecraft	Year of encounter	Ref. to geology[b]	Ref. to crater density[b]
Phobos	Mariner 9	1971	1	–
	Viking 1	1977	2	2
	Phobos 2	1988	3	–
	Mars Global Surveyor	1999	4	–
Diemos	Viking 1	1977	2	2
			5	
Gaspra	Galileo	1991	6	7
Ida	Galileo	1993	8	9
Mathilde	NEAR Shoemaker	1997	10	11
Eros	NEAR Shoemaker	2000	12	13

[a] Or presumed captured asteroids in the case of the martian moons Phobos and Diemos.
[b] References: 1, Masursky *et al.* (1972); 2, Veverka and Duxbury (1977); 3, Murchie and Erard (1996); 4, Thomas *et al.* (1996); 5, Thomas *et al.* (2000); 6, Carr *et al.* (1994); 7, Chapman *et al.* (1996b); 8, Sullivan *et al.* (1996); 9, Chapman *et al.* (1996a); 10, Thomas *et al.* (1999); 11, Chapman *et al.* (1999); 12, Robinson *et al.* (2002); 13, Chapman *et al.* (2002), Veverka *et al.* (2001).

currents induced by solar wind (Sonett, 1979) are still discussed, and these would provide heat externally not give a totally internally heated body. Thus it would be possible for a body to be CI-like throughout (i.e. containing 10–20 vol. % water) and dried out only in the outer megaregolith layers. Such a model would be consistent with asteroid densities, of course, but seems never to have been proposed.

2.2 Impact and cratering processes

2.2.1 The geology of asteroids

To date, six asteroids have been visited by robotic spacecraft (Table 2.5), and more will be visited in the next few years. It is important to realize how recently this has happened, and how little information was available – beside astronomical spectra and dynamical information – before these encounters. Most of the ideas in the literature about asteroid parent bodies are derived from meteorite data and predate these images and the fundamental information they contain. Thus many of our ideas need to be revisited.

Figure 2.10 shows a collection of images of asteroids taken by robotic spacecraft. The Galileo spacecraft took pictures of the asteroids Gaspra and Ida on its passage

2.2 Impact and cratering processes

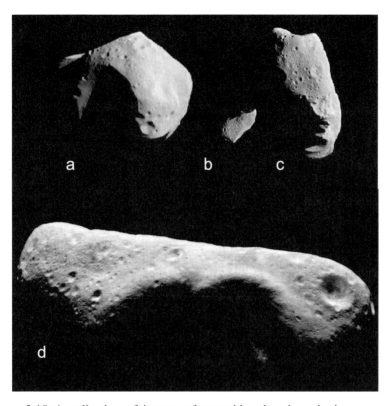

Figure 2.10 A collection of images of asteroids taken by robotic spacecraft. (a) **Mathilde**, as imaged by the NEAR Shoemaker spacecraft. (b) **Gaspra** and (c) **Ida**, both imaged by the Galileo spacecraft. The remaining images of Eros are all from the NEAR Shoemaker spacecraft. (d) The southern hemisphere of **Eros** (33 km). The site at which the spacecraft finally landed is just into the shadows, slightly left of center. (e) **Grooves** on Eros from an orbital altitude of 51 km cutting across the oldest, most subdued craters but underlying the smaller, younger craters. (f) An **ancient ridge** on Eros' southern hemisphere, southwest of a large, 5.3 km crater. The large number of superimposed impact craters suggests that the ridge is among the older features on Eros. Image ∼11 km top to bottom. (g) Evidence for abundant **regolith** on Eros. Unlike the Moon's surface, which is dominated on all scales by craters, the surface of Eros is dominated by a blanket of regolith. Boulders litter the landscape, the smallest craters are obscured, and many of the low spots are extremely flat, and appear infilled. The whole scene is ∼1.1 km. (h) **Regolith and boulders.** The four large boulders in the lower left are ∼90 m across. In the smooth patch at upper right, the regolith fills a depression and in places regolith appears to have flowed. (i) One of the three large craters on Eros, **crater Psyche (5.3 km)**, showing several troughs and scarps perhaps resulting from a large impact elsewhere on the asteroid, the crater's raised rim, bright patterns on the crater wall from dark material moving downslope, and a large boulder perched on the crater wall (because of its elongated shape the gravity "lows" on Eros are not necessarily in the lowest parts of craters). (j) **Boulder-filled, concave depression** at the southwestern edge of the saddle-shaped Himeros from an orbital altitude of 52 km. (k) One of the **final Eros images** as the spacecraft descended to the surface. The image is 12 m across. The cluster of rocks at the upper right measures 1.4 m across. All photographs courtesy of APL/NASA.

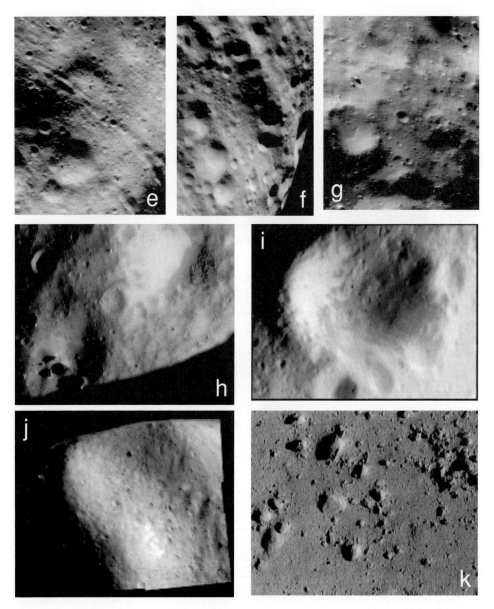

Figure 2.10 (cont.)

through the Asteroid Belt en route to Jupiter. The geology of these objects has been summarized by Belton *et al.* (1992) and Sullivan *et al.* (1996), respectively. Ida has a moon, Dactyl, that has been described by Chapman *et al.* (1995). Mars-orbiters have also taken images of the moons of Mars, which are thought to be captured asteroids, Diemos (Veverka and Thomas, 1979) and Phobos (Thomas *et al.*, 1996). NEAR Shoemaker took images of Mathilde on its way to Eros (Thomas *et al.*,

1999). However, by far the best documented asteroid is the near-Earth asteroid Eros, onto which the NEAR Shoemaker spacecraft landed after orbiting for a year (Fig. 2.10d). While these asteroids have much in common, what is also notable are their differences. Chapman (2001) has compared spacecraft images of asteroids. Gaspra has essentially no large craters and its small craters are well below saturation densities, suggesting that either the object is very young or is made of strong metal. Ida is saturated with craters (or nearly so) with a ~2 Ga megaregolith. Ida might be a rubble pile consisting of two large pieces. Mathilde is like Ida in many respects, but supersaturated with large craters. Mathilde's low density suggests large numbers of voids that would give the asteroid unusual physical properties with respect to shock and impact ejecta. Eros is like Ida but appears to be a shattered shard of a larger object. In addition, Deep Space 1 has flown past Braille and Stardust has flown past Annefrank, but the images are of much lower quality than the others.

Eros shows a variety of surface features and evidence for substantial regolith (Robinson *et al.*, 2002; Thomas *et al.*, 2002). There is at least one global planar structure that is indicated by large-scale facets, grooves, and ridges. The asteroid appears to be substantially fractured. There are a great many impact craters with depth-to-diameter ratios similar to those on the Moon (~0.2), especially when fresh. There are three especially large craters, Himeros (11 km), Shoemaker (7 km), and Psyche (5.3 km), Shoemaker having spread ejecta blocks across the asteroid. There is no evidence for strata within the craters and <200 m craters are deficient, having been buried by regolith. There is evidence for mobility of the regolith.

2.2.2 Craters

All of the asteroids imaged by robotic spacecraft to date, and this includes the moons of Mars that are assumed to be captured asteroids, have cratered surfaces. The mechanism of cratering, at least on Earth and the Moon, has been thoroughly reviewed (Melosh, 1989) but very little is known about cratering on asteroids where it may be expected to be quite different. For instance, the lower gravity fields will mean that transported volumes will be greater. There is also considerable uncertainty about the interior structure of asteroids, and questions as to whether the gravity field or the strength of the asteroid will prevail in determining the physics of interaction. Theoretical treatments exist for both options. Crater densities are at saturation on all the asteroids observed by spacecraft, but for Eros, where the best data exist, small craters are absent. It is presumed that they are buried under the regolith. The three major craters on Eros were mentioned above. Sources of crater density information are listed in Table 2.5.

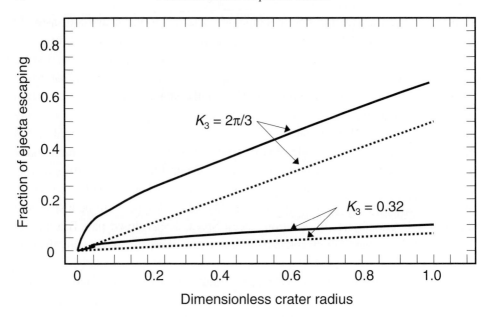

Figure 2.11 Fraction of ejecta escaping as a function of location in the crater from which ejection occurs (Asphaug and Nolan, 1992). One of the surprises of the spacecraft images of asteroids is that the surface appears to be covered with a thick layer of unconsolidated surface material called regolith. It had been assumed that the gravity of the asteroid was too small to retain such material. However, new calculations suggest otherwise. Solid lines are for rock, broken lines are for sand. The volume fraction of ejecta escaping is $3K_3/2\pi$ $(r/2)$ $s/2$, where K_3 and s are constants and r is the ratio of crater size to target asteroid size. The curves with K_3 values of $2\pi/3$ are theoretical upper limits; the curves with K_3 values of 0.32 are based on experimental data and more realistic. For material from most of the crater, probably less than about 10% escapes from the asteroid as a result of the impact.

2.2.3 Physical and chemical processes accompanying surface working

On the Moon, impacts capable of producing craters up to about 500 m in diameter dominate in the process of comminuting bedrock into an unconsolidated surface layer of particles, comparable in size to the grains in the precursor rocks, termed "regolith" (Hörz and Cintala, 1997). The production rate of regolith on the Moon has decreased (from a few millimeters per year ~3 Ga ago to <1 mm/year at present) as the thickness of the regolith built up and it became harder for impacts to penetrate to bedrock. This comminution process affects some minerals more than others so that feldspar is enriched in the finer fraction. This might explain why agglutinates have feldspathic composition. Until we have brought back samples from an asteroid regolith, we can only speculate and extrapolate using ideas from the Moon. This is difficult because the gravitational field of the asteroid is much less, impact velocities

Table 2.6 *Regoliths on asteroids (Asphaug and Nolan, 1992).*[a]

Property	Crater radius (km)		
	2	4	6
R_c/R_t[b]	0.31	0.62	0.93
Projectile radius (code) (m)	76	150	284
Ejecta escaping (code) (%)	30	85	99
Ejecta escaping (scaling) (%)	28	180	280
Transport distance (code) (m)	100	500	Escapes
Final state (code)	Intact	Rubble	Rubble

[a] Calculations for impacts into a 6.4 km basalt asteroid at 5.3 km/s. Two methods were used, scaling and hydrocode computations.
[b] Scaled crater radius, i.e. radius of the crater divided by radius of the target asteroid.

are lower, and target compositions more volatile. The gas-rich regolith breccia meteorites described above provide some information, but they are now lithified rocks and have lost many of their regolith properties. What information can be gleaned, however, also suggests significant differences in regolith processes on asteroids and the Moon (McKay *et al.*, 1989). One of the major issues used to be whether asteroids have a regolith at all, but this seems to have been resolved by spacecraft images. Now, the main question is determination of the thickness of regolith and how it varies with topography.

2.2.4 Regolith depth

Early theoretical estimates of regolith thickness by Langevin and Maurette (1980) suggested that weak asteroid-sized bodies were capable of retaining little ejecta (<100 m), with relatively little dependence on the size of the asteroid, while strong bodies larger than about 50 km would have regoliths of 100–900 m depending on asteroid size. Housen *et al.* (1979) calculated that a 10 km asteroid should have no regolith, but 100, 300, and 500 km asteroids should have regoliths of 0.2, 3.5, and 1.2 km, respectively. However, these calculations are strongly dependent on the ejecta velocities, which, in turn, depend on target strength. Housen (1992) and Asphaug and Nolan (1992) have presented various semi-theoretical treatments to indicate that even asteroids in the 10 km size range are capable of retaining substantial regoliths if they have low target strength. Examples of these calculations are shown in Fig. 2.11 and Table 2.6, which were prompted by Galileo's observation of a significant regolith on Gaspra (with dimensions 19 × 12 × 11 km). Scheeres *et al.*

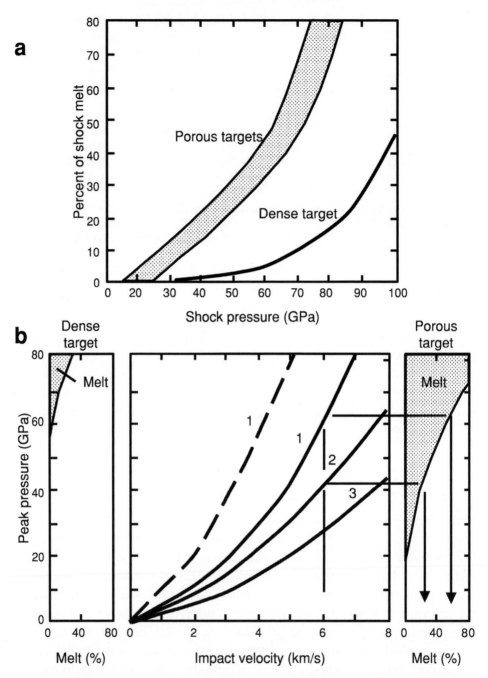

Figure 2.12 Impact and melt production as described by Hörz and Schaal (1981). (a) Shock pressures are lower and melt production higher for impact of dense basalt (actually Apollo 17 sample 75035) into targets of ~35% porosity than they are for dense targets, regardless of grain size. (b) Peak pressure as a function of impact velocity for various impactors into dense basalt (broken line) and porous

(2002) review this topic and the surprise discovery of the importance of thick dynamic regoliths on asteroids.

Besides impact ejection, the smallest asteroids are often assumed to lose regolith by centrifugal forces as a result of their high rotation rates. This is because there is a rather strong relationship between asteroid size and rotation rate, with asteroids smaller than 200 m spinning faster than 2.2 times per hour while larger asteroids rotate typically a few times per hour to a few tens of times per hour. Asteroid size (D) is determined from the absolute magnitude (H) of the asteroid using a relationship that involves albedo, such as:

$$\log D = 0.2085H + 1.3725 \quad \text{for albedo} = 0.25 \tag{2a}$$

$$\log D = 0.2085H + 1.7285 \quad \text{for albedo} = 0.05 \tag{2b}$$

2.2.5 Amount of melt produced on impact

In addition to causing loss of ejecta, impact can also result in the presence of impact melts on the surfaces of asteroids and these are important since they are sometimes observed in meteorites. Some authors have suggested that chondrules are crystallized impact-melt spherules. About two decades ago it was believed that impact velocities on asteroids were too low for significant amounts of melt production, but recent works suggest this might not be true. Figure 2.12 shows some work of Hörz and Schaal (1981) that describes impact and melt production on porous targets compared to dense targets. Because the rate of deposition of energy in a porous target is much higher than for a dense target that can more effectively transmit energy, the amount of melt produced is very much higher. A pressure of 60 GPa (600 kbar) can produce as much as 50 vol. % melt in the porous target, compared with only 5 vol. % in a dense target. In fact, iron, dense, or porous basalt impactors into a porous surface will produce significant melt; a dense projectile impacting at 5 km/s – a typical impact velocity for the Asteroid Belt a few 10^5 years after Solar System formation – gives a pressure of 60 GPa and ~50 vol. % melt production.

---

basalt (solid lines). The impactors are iron (1), dense basalt (2), and porous basalt (3). (Dense basalt impacting dense basalt follows the same curve as iron impacting porous basalt.) Note that while little or no melt is produced by impact into dense targets, a typical stony meteorite impacting into loosely consolidated regolith at ~6 km/s would produce ~20% melt while an iron meteorite would produce ~60% melt, according to these results.

3

Chondrites and their main properties

3.1 Classification and composition

3.1.1 Chondrite classes

The chondrite classes that were introduced in Chapter 1 are the enstatite chondrites, at the reduced end of the spectrum and with large metal–silicate fractionation between the so-called EH and EL chondrites. There are many published reviews of these meteorites (Keil, 1968; 1989; Sears *et al.*, 1982b; Zhang *et al.*, 1995). The ordinary chondrites appear in the middle of the Urey–Craig plot with metal–silicate fractionation separating the H, L, and LL chondrites. Urey and Craig (1953), Keil and Fredriksson (1964), and Fredriksson *et al.* (1968) discussed these classes. The R (after Rumuruti) chondrites are superficially similar to the ordinary chondrites although they contain considerably more matrix and have unusual oxygen isotope properties (Weisberg *et al.*, 1991; Kallemeyn *et al.*, 1996). The carbonaceous chondrites contain little metallic iron, although some contain sulfides and all contain considerable amounts of iron in the form of clay minerals and magnetite. The carbonaceous chondrites can be subdivided into CO chondrites (McSween, 1977a; Scott and Jones, 1990; Sears *et al.*, 1991), CV chondrites (Van Schmus and Hayes, 1974; McSween, 1977c), CR chondrites (Bischoff *et al.*, 1993b; Weisberg *et al.*, 1993), CK chondrites (Kallemeyn *et al.*, 1991), CM chondrites (McSween, 1979a), and CI chondrites (McSween, 1979b). The carbonaceous chondrites contain little metal and have suffered little metal–silicate fractionation, however, redox state varies. The CO and some CV chondrites are somewhat more reduced than the other C chondrites. Thus the CV chondrites are subdivided into reduced and oxidized subgroups. The CI and CM chondrites contain small amounts of organic materials that have been studied intensely because of possible relevance to the origin of life in the Solar System (Cronin *et al.*, 1988). In addition to these groups, there are the metal-rich chondrites like Allan Hills 85085 that are referred to as CH chondrites (Grossman *et al.*, 1988b; Scott, 1988; Weisberg *et al.*, 1988; Bischoff *et al.*, 1993b).

There are many chondrites that do not belong to any of these groups, such as those resembling Kakangari (Weisberg *et al.*, 1996).

It is probably best to consider the "primitive achondrites," the brachinites (Nehru *et al.*, 1992), and the acapulcoites (Palme *et al.*, 1981; McCoy *et al.*, 1992) as chondrite classes. These are meteorites without chondritic textures but with chondritic bulk composition and it is assumed that they are chondrites that melted without change in composition.

Finally, the interplanetary dust particles are chondritic in composition and represent an additional source of chondritic material (Bradley *et al.*, 1988; Brownlee *et al.*, 1997).

Iron meteorites are beyond the scope of the present work, but several iron meteorite classes have been linked to chondrite classes. Some iron meteorites contain chondrules (Olsen and Jarosewich, 1971), while some iron meteorite classes have been linked to chondrite classes by geochemical trends (Scott and Wasson, 1975). Buchwald (1975) provides the best overview of iron meteorites and their properties, particularly their structures.

3.1.2 Bulk compositional trends within the chondrite groups

The behavior of iron and oxygen in chondritic meteorites is of such fundamental importance that they are the major determinant of the chondrite classes. In turn, the chondrite classes are thought to carry important information about the early Solar System and processes occurring there. Prior (1916) suggested that the oxidization trends he had observed, and summarized as "Prior's Laws," were the result of the production of meteorites sequentially in a magma chamber. Urey and Craig (1953) realized that this could not be the case because, as apparent from Fig. 1.2, the texture of these meteorites is clearly not igneous. Such a variety of materials and range of textures and sizes of components do not occur in igneous rocks, they argued. At a loss for a process that could have occurred on the parent body, Urey and Craig (1953) suggested that perhaps these separations of metal and silicates occurred in the nebula. We will return to these issues later, but for the moment these things are mentioned because they underscore the importance of iron and oxygen trends in chondritic materials. But iron and oxygen are not the only elements present. What can be learned from the other elements?

The major sources of bulk compositional data for chondrite meteorites are the compilations of wet chemical data by Wiik (1969) and Jarosewich (1990a). While the X-ray fluorescence data of Ahrens and his co-workers (Ahrens, 1965; 1970; Ahrens and Von Michaelis, 1969; Von Michaelis *et al.*, 1969a; 1969b), isotope dilution data (Nakamura and Masuda, 1973; Nakamura, 1974), and INAA (instrumental neutron activation analysis) data (Loveland *et al.*, 1969; Schmitt *et al.*, 1972;

Laul *et al.*, 1973; Anders *et al.*, 1976; Matza and Lipschutz, 1977; Takahashi *et al.*, 1978a; 1978b; Wolf *et al.*, 1980; Kallemeyn and Wasson, 1982a; 1982b; Sears *et al.*, 1982a; Kallemeyn *et al.*, 1991; 1994) are also very important for bulk element determination, they do not determine oxygen and the distribution of iron between phases. Noble gases in chondrites have been reviewed by Swindle (1988).

When discussing chemical trends in rocks that are essentially solar in composition, it is necessary to generalize. There are about 90 elements in chondrites and at some level of detail all differ in their properties. However, there are few physical phases for the elements to be located and, to a reasonable approximation, their phase determines their behavior as the meteorites undergo a number of processes in their history (Larimer, 1988). So we can collectively discuss the elements that tend to concentrate in the silicate phase (Si, Mg, Na, Al, Ca, etc.) as "lithophiles" (stone-loving), the metal phase (Fe, Ni, Co, Ir, etc.) as "siderophiles" (metal-loving), the sulfide phase (Cu, In, Tl) as "chalcophiles" (sulfide-loving), and the gas phase (Ar, Ne, Xe, N, etc.) as "atmophiles" (air-loving). It is also helpful to distinguish between highly volatile, moderately volatile, and refractory elements (e.g. Lipschutz and Woolum, 1988; Palme *et al.*, 1988). Some elements change their behavior depending on the situation, and some show mixed behavior, the important example being Fe that has siderophile, lithophile, and chalcophile tendencies. On the one hand, this complicates an interpretation of Fe abundances, but on the other hand it makes it possible to set up useful equations relating composition to physical conditions. For example, when the distribution of iron between the silicate and metallic phases is known, it is possible to calculate the "equilibrium temperature" at which the system formed.

The FeO content can be determined by the crucial redox reaction:

$$Fe(g) + H_2O(g) = FeO_{l \text{ in soln}} + H_2(g) \tag{3}$$

for which:

$$\ln K = \ln[a(FeO)P(H_2)/P(Fe) P(H_2O)] \tag{4}$$

where $a(FeO)$ is the activity of FeO. Since:

$$\Delta G = -RT\ln K = \Delta H - T\Delta S \tag{5}$$

$$-\Delta H/RT + \Delta S/R = \ln[a(FeO)P(H_2)/P(Fe)P(H_2O)] \tag{6}$$

By Dalton's Law of Partial Pressures we can write:

$$P(Fe) = P(H_2)A(Fe)/2A(H) \tag{7}$$

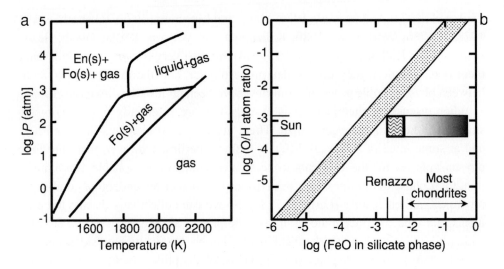
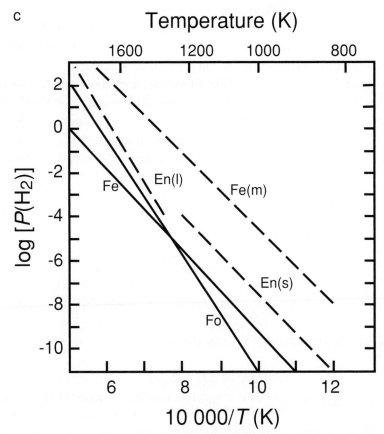

3.1 Classification and composition

Where the A terms refer to cosmic abundances. Then by inserting appropriate values for ΔH, ΔS, $A(\text{Fe})$, $A(\text{H})$ and assuming $a(\text{Fe}) \approx X_{\text{Fa}}$ equation (6) becomes:

$$\log\left[\frac{P(\text{H}_2\text{O})}{P(\text{H}_2)}\right] = \log[X_{\text{Fa}}] - \frac{1837}{T} + 1.64 \qquad (8)$$

The solution of this equation is shown in Fig. 3.1. From this figure it can be seen that the low FeO content of olivine in low-FeO chondrules (such as those in Renazzo) is close (but not equal) to that expected for the solar $\text{H}_2/\text{H}_2\text{O}$ ratio (e.g. Wood, 1967b; Johnson, 1986; Kring, 1986; Palme and Fegley, 1987) but that the high FeO of most chondrules is about two orders of magnitude higher than predicted from equation (8) and cosmic abundances (Suess and Thompson, 1983). Oxidation–reduction relations shown by ordinary chondrites have been discussed by many authors (Müller et al., 1971; Rubin, 1980; Rubin et al., 1988).

Trends in elemental abundance are best represented on a volatility plot of the sort first used by Larimer and Anders (1967). Strongly correlating elements appear as horizontal trends on this plot and by sorting the elements according to calculated condensation temperature, volatility trends can be identified as downward slopes (Fig. 3.2). The first geochemical trend to look for is the siderophile element pattern that was first observed by Urey and Craig (1953) in Fe. This is the trend that gives rise to to the H, L, and LL chondrites. It can be seen that EH chondrites have higher siderophile element abundances than EL chondrites. The variation in siderophile element abundance in H, L, and LL is qualitatively similar to that of Fe in H, L, and LL, but the effect is greater for the more refractory (less volatile) elements

Figure 3.1 Phase diagrams for magnesium–iron silicates in a solar gas. (a) At high temperatures and low pressures the system is entirely gaseous and as the temperature drops forsterite (Fo) and then enstatite (En) form. At high pressures and temperatures, the silicates are liquids. Thus, in order to produce chondrules in a gas of cosmic composition, whether by melting or condensation, pressures must be very high ($\geq 10^3$ atmospheres). (b) Another difficulty with producing chondrules in a gas of solar composition is that over a very wide range of conditions the chondrules, even the highly reduced chondrules from the Renazzo chondrite, are much too oxidized. The composition of silicates in Renazzo, and even more so the silicates in other chondrites, plot to the right of the theoretical diagonal band. (Figures adapted from Wood, 1962.) (c) Pressures and temperatures at which solid phases form in the nebula with (broken lines) and without (solid lines) supercooling. Without supercooling, metallic Fe and the silicates at similar temperatures and pressures are Fe-free. However, with supercooling, iron nucleates much less readily than silicates so its condensation is suppressed to lower temperatures. Thus the amount of Fe(g) in the gas is increased and more Fe enters the silicates. In this way, supercooling produces silicates with higher Fe than at equilibrium (Blander and Katz, 1967).

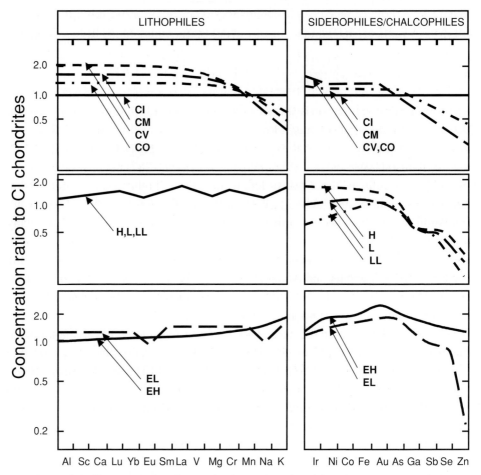

Figure 3.2 Elemental abundances in the major chondrite classes expressed as ratios to CI chondrite abundances and plotted in order of increasing volatility (as determined by thermodynamic calculations for a system with solar bulk composition) (Sears and Dodd, 1988). Refractory and moderately volatile elements show class-to-class variations that could either reflect differences in precursor materials or differences in chondrule (and CAI) type and abundance. Siderophile elements show variations associated primarily with the metal–silicate fractionation. Not discussed at length in this work are the highly volatile element depletions that are related to petrographic type and therefore metamorphism.

plotting to the left. For the C chondrites the trend is CO, CV > CM > CI for the least volatile elements and the reverse for the most volatile elements. In short, "siderophile element fractionation" seems to be displayed in a qualitatively similar fashion by most chondrite classes, and appears strongest for the most reduced classes. This is commonly attributed to a separation of metal and silicate and is referred to as "metal–silicate fractionation."

Much of the fine detail in Fig. 3.2 can be attributed to volatility differences or to differences in degree of siderophile behavior. For example, Fe shows a smaller range of values in the H, L, and LL chondrites than, say, Ir, because much of the Fe is in phases other than metal.

The cause of the abundance patterns in the so-called moderately volatile lithophile elements has resulted in considerable debate. Larimer and Anders (1967) suggested that the horizontal parts of Fig. 3.2 (they called them "plateaux") are caused by the mixing of two components, volatile-rich matrix with volatile-poor chondrules. Thus the abundances of moderately volatile elements in the H, L, and LL chondrites represent a mixture which is ~25% chondrules and 75% matrix. Each class, Larimer and Anders argued, contains horizontals (see Fig. 3.2) that are the result of this two-component mixing where the chondrules, metal, and sulfide grains are depleted in volatile elements while the matrix has its full cosmic proportions of volatile elements. The volatile components had been cycled through a high-temperature event, namely chondrule formation, and the other component, the matrix, had escaped. Wasson and others argued that this was not the case, but that instead there were no steps, just a steady slope downwards on Fig. 3.2, that was caused by processes that predated the chondrule-forming process. Presumably we were seeing the very process by which elements in the gaseous primordial nebula condensed to produce solid materials. The missing volatiles were still in the gas when the solids formed and were later swept away by the Sun. Cassen (2001) has tried to relate the moderately volatile element abundance patterns to nebular processes.

It is possible to calculate volatility in the primordial solar nebula by thermodynamic calculations. By using chemical equilibrium relationships and mass balance, and Dalton's Law of Partial Pressures, the temperature at which elements condense can be calculated so that we can plot the elements in order of decreasing condensation temperature (Fig. 3.2).

The most highly volatile elements, such as In, Tl, Bi, and Pb are clearly due to processes other than metal–silicate fractionation. This is because they vary in abundance in a manner dependent on petrographic type. They increase in abundance as petrographic type decreases and the effects are not small; these elements can be 10 000-fold more abundant in type 3 ordinary chondrites than they are in type 6. Here too there is uncertainty over interpretation. One possibility is that these highly volatile elements were driven off by metamorphism, the more intense the metamorphism the greater the loss of the elements. Heating experiments in the laboratory are consistent with this interpretation in the case of some chondrite classes but not others. Another possibility is that we are seeing the condensation process at work (Larimer and Anders, 1967). Thermodynamic calculations show that these elements condense over a relatively limited temperature interval so that

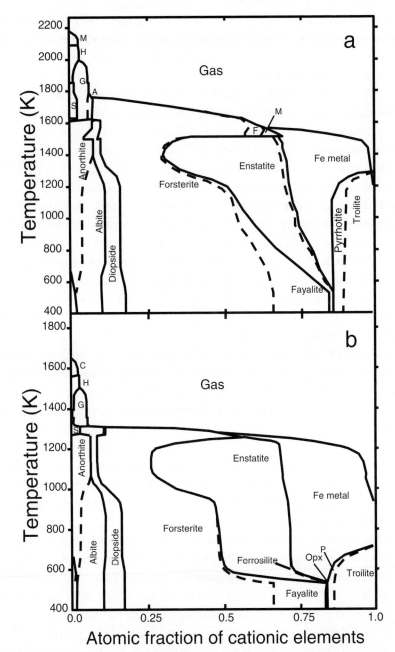

Figure 3.3 Equilibrium thermodynamic calculations for the solid phases present as a function of temperature. The expected effects of both condensation from a homogeneous gas during a decrease in temperature and evaporation of solids during an increase in temperature can be estimated from these diagrams. (a) A gas of solar composition except that the proportion of elements that form solids has been increased 1000-fold to represent dust-to-gas enhancement. (b) A gas of solar

3.1 Classification and composition

small differences in temperature, less than 100 °C produce 10 000-fold differences in abundance. In this way it is actually possible to determine accretion temperatures for each class of meteorites that are in the range 400–600 K. However, these efforts have been criticized because of their heavy dependence on poorly known thermodynamic quantities (Blander, 1975).

It has long been realized that many properties of chondrites can be understood in terms of equilibrium condensation–evaporation calculations (Lord, 1965; Larimer, 1967; Grossman, 1972; Sears, 1978); The most recent of such calculations are shown in Fig. 3.3 (Wood and Hashimoto, 1993).

At high temperatures a number of minerals rich in refractory elements like Ca, Al, and Ti are produced, but the cosmic abundance of these is fairly low and the amount of solid material is small. At lower temperatures the major elements (Mg, Si, and Fe) form solids and the silicates (enstatite and forsterite) and metal form. At slightly lower temperatures silicates containing moderately volatile elements like Na and K are stable. At the lowest temperatures, iron sulfide forms and iron is oxidized and takes up solid solution in the silicates. At the very lowest temperatures, not shown in Fig. 3.3, water becomes a stable solid in the form of ice and chemically bound water. It is possible to modify these calculations to allow for increases in the dust-to-gas ratio, in dusty regions of the nebula or in the vicinity of newly forming planetesimals. The results of such calculations are also shown in Fig. 3.3. Because the gas is mainly H_2, increasing the amount of dust (by a factor of 1000 in this example) increases the equilibration temperatures and makes conditions less reducing – stability of oxidized Fe increases and the stability of metallic Fe decreases.

What is not clear from equilibrium calculations like those shown in Fig. 3.3 is whether this is a condensation or evaporation sequence. It is often interpreted as a condensation sequence, but equilibrium could equally well be approached from lower temperatures as solids were heated and the volatile elements evaporated. The chondrules clearly provide evidence for a heating event, and the temperatures involved in chondrule formation were high enough to cause evaporative loss of

---

composition. The refractory (Ca- and Al-rich) phases have been abbreviated as follows: C, corundum (Al_2O_3); H, hibonite ($CaAl_{12}O_{19}$); G, gehlenite ($Ca_2Al_2SiO_7$); S, spinel (Mg_3AlO_4). M stands for melt; P, pyrrhotite (FeS_x); Opx, orthopyroxene (Wood and Hashimoto, 1988; 1993). The main effect of the increased dust-to-gas ratio is to increase the temperatures at which solids are stable but since the "gases" are mostly hydrogen, oxygen-bearing solids (like fayalite, Fe_2SiO_4, and ferrosilite, $FeSiO_3$) become more stable and metallic iron less stable. Several trace phases included by Wood and Hashimoto (1993) whose fields are too small to show on this figure have been omitted.

volatile elements if the kinetics were favorable. Thus whether chondrules were open or closed systems during their formation has wider implications than just the history of the chondrules. It could explain the bulk composition of the chondrites rather than appealing to nebular processes for which we only have speculations.

3.1.3 Textural differences between the classes

In addition to the mineralogical and bulk compositional properties mentioned above, each chondrite class displays an almost unique combination of metal and chondrule sizes and matrix abundance (Figs. 1.8 and 1.9).

Chondrule sizes and shapes have been explored by many authors (Martin et al., 1975; Martin and Mills, 1976; 1978; 1980; Hughes, 1978; King and King, 1978; 1979; Zbik and Lang, 1983; Rubin and Keil, 1984; Rubin and Grossman, 1987). To a good approximation chondrules are spherical, but occasionally a meteorite is discovered in which flattening is observed and ascribed to an encumbrant load or shock. The shape of the chondrule size distribution plot depends on the production mechanism for the chondrules, fragmentation producing a particular kind of distribution. However, the results have been indecisive. What is clear, however, and worth stressing, is that each meteorite class has a particular distribution of chondrule sizes almost to the point that this could be a classification parameter. The same is true of metal grain sizes.

The amount of matrix decreases as the amount of volatile components decreases, which is consistent with the observation that the volatiles are located in the matrix.

3.1.4 Oxygen isotope differences between the classes

Oxygen is an important element because it is abundant, is present in a large number of minerals, and is one of the few elements that can be simultaneously located in gas, liquid, and solid phases in many astronomical environments. Oxygen also has three stable isotopes, ^{16}O, ^{17}O, ^{18}O, so it is possible to distinguish mass-dependent and mass-independent processes. Oxygen isotope abundances proved to be a useful means of classifying chondritic meteorites even though the cause of the isotope variations is completely unknown. The data for chondrites and chondrules have been published a great many times (Clayton et al., 1981; 1983; 1987; 1991; 1992; Gooding et al., 1982; 1983; Clayton, 1983; 1993; Clayton and Mayeda, 1984; 1985; McSween, 1985). When plotted on a three-isotope diagram (Fig. 3.4), most of the chondrite classes plot on trend lines or as discrete fields, the ordinary chondrites above the terrestrial line, the carbonaceous chondrites (CI excepted) below the terrestrial line, and the enstatite chondrites plot on the terrestrial line. CV and CM

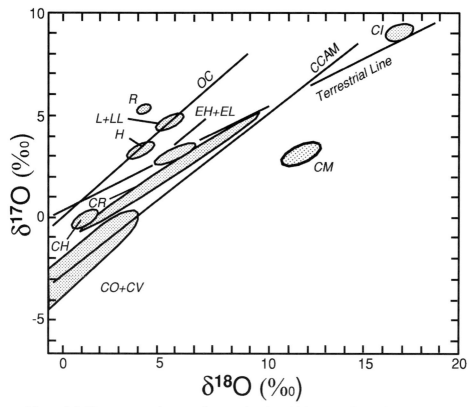

Figure 3.4 Three oxygen isotope diagram for chondritic meteorites (after Clayton et al., 1991). The δ quantity refers to difference in parts per thousand (‰) between the $^{17}O/^{16}O$ or $^{18}O/^{16}O$ ratio of the sample and a standard (Standard Mean Ocean Water, SMOW), i.e. $\{[(^{18}O/^{16}O)_{sample}/(^{18}O/^{16}O)_{standard}] - 1\} \times 1000$, and similarly for $^{17}O/^{16}O$. The chondrite classes plot in discrete fields or trend lines: ordinary chondrites plot above the terrestrial fractionation line (slope 0.5); the carbonaceous chondrites plot below the line (CI excepted); the refractory inclusions plot on a slope 1 (mixing) line; and the enstatite and CI chondrites plot on the terrestrial line.

chondrites plot on a line with slope ∼1, as do the H, L, and LL chondrites, the two lines that approximate slope 1 intercepting at a $\delta^{18}O$ and $\delta^{17}O$ value of −40‰.

These data are also frequently used for petrogenesis. The two slope ∼1 lines could represent mixing lines between, say, solids with a $\delta^{18}O$ and $\delta^{17}O$ value of −40‰ and gases containing heavy oxygen, the gases being any ambient gas and possibly even the nebula gases. However, laboratory experiments have produced slope 1 lines on the three-isotope plot, suggesting that fairly straightforward physical processes could produce such lines (Thiemens, 1988; 1996). One of the major objectives of the Genesis mission is to measure the oxygen isotope ratios for the solar wind oxygen. Then at least we will presumably have an "origin" point to plot on Fig. 3.4.

3.2 Formation history

3.2.1 Accretion, agglomeration, brecciation, and final assembly

When Sorby (1864) first saw chondrites through a microscope he likened them to volcanic rocks noting particularly the "globules" that had "cystallized after solidification," the abundance of glass and gas cavities in the silicates that resemble those found in terrestrial lavas. He later repeated these arguments and likened the overall texture of chondrites to "consolidated tuff or ashes." Tschermak (1885) remarked on the "tuff like" character of meteorites while Haidinger (1867) considered chondrules as abraded rock fragments of the sort found in volcanic eruptions and explosions. Berwerth (1901) suggested that chondrite stones were tuffs metamorphosed by heat. Borgstrom (1904) argued that the intergrown texture of chondrules and matrix suggested *in situ* crystallization. Thus many of the steps in the evolution of these rocks recognized by modern meteorite researchers were identified early in their study. We would not refer to "accretion," "agglomeration," "brecciation," and "final assembly" and we consider them different processes, occurring at different times and under different physical conditions.

By "accretion" we mean nebula dust coming to form planetesimals, or nebula dust being collected on the surface of planetesimals. By "agglomeration" we mean this material forming the surface rocks on the planetesimals and by "brecciation" we mean disruption of this rocky surface by impact and reaccumulation of the fragments. "Final assembly" means taking these materials and turning them into the rocks we see today. Somewhere during the whole process, the components were made and secondary processes occurred in addition to impact – these were aqueous alteration and metamorphism. Meteorites often show evidence for these kinds of secondary process. What these four processes have in common is that they all deal with bringing the component parts of the meteorite together to create its final texture and composition. Since all we have to go on is the final rock, it is difficult to distinguish these processes and petrographic observations are often open to multiple interpretations, especially since little is known about the individual processes. In particular, a considerable amount of time has been spent trying to identify properties that reflect "nebula" as opposed to "planetary" properties. It is a very complicated process with any number of variations. For example, if the planetesimal was water-rich, like a CI chondrite, impact could produce a temporary atmosphere whose effects in the meteorite could be mistaken for nebula effects.

The subject of agglomeration and final assembly of meteorites has been discussed from many perspectives. Solar wind tracks and gases in grains in primitive meteorites provide meteorite evidence for an active early Sun (Caffee *et al.*, 1987; 1988). This has been the topic of a workshop and book (Pepin *et al.*, 1980). Such properties enable compaction ages – the ages at which components came together – to be determined. Grain-to-grain differences in charged particle tracks due to fission of ^{244}Pu suggest that carbonaceous chondrites agglomerated within a 10^8 year period (Macdougall and Kothari, 1976; Caffee and Macdougall, 1988).

Macroscopic brecciation has been reviewed by Bunch and Rajan (1988). Keil (1982) has defined six types of breccias: breccias with lithic fragments, regolith breccias, fragmental breccias, impact melt breccias, granulitic breccias, and primitive breccias. These terms are generally self-explanatory, but granulitic breccias are metamorphosed fragmental breccias and primitive breccias are type 3 ordinary chondrites. Several authors have pointed out that xenolithic clasts (clasts of a different chemical class to the host meteorite) are very rare in breccias but when they do occur they are usually CM chondrites. Gas-rich regolithic breccias are very important and are discussed elsewhere.

A possibly unique means of exploring accretion mechanisms as well as parent body properties is the magnetic properties of meteorites. Magnetic studies of chondrites have been performed by many authors (Lanoix *et al.*, 1977; 1978; Sugiura *et al.*, 1979; Funaki *et al.*, 1981; Nagata and Funaki, 1983; Sugiura and Strangway, 1985; 1988; Stepinski and Reyes-Ruiz, 1993), especially in terms of the magnetic field of the primordial solar nebula. Sugiura and Strangway (1988) have reviewed the topic. The most important message in the meteorites is that the field they have experienced is much greater than might be predicted and that either the interplanetary magnetic field was once much greater than it is now or the field strength was intensified somehow as the meteorites cooled through their Curie points. It has been suggested that during its τ-Tauri phase the Sun might have generated a field strong enough to explain meteorite magnetism or there might have been enhancement of the field by shock. Clasts and other components within brecciated meteorites generally have randomly oriented fields, which suggests magnetization before brecciation.

3.2.2 Metamorphism and aqueous alteration

The textures of chondrites provide evidence for them having experienced various degrees of alteration by metamorphism and hence to their sorting into types 3–6. It is possible to estimate metamorphic temperatures that correspond to these types, but the estimates are fairly approximate. Probably the best constraints on metamorphic temperatures are the behavior of metal and sulfides, which melt at 950–1000 °C and form kamacite and taenite phases above $\sim$350 °C. The partitioning of Fe and Mg between coexisting orthopyroxene and diopside and of Ca between coexisting pyroxenes can also be used to estimate equilibration temperatures (Ishii *et al.*,

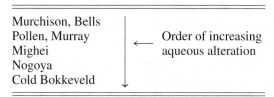

Table 3.2 *Aqueous alteration of CM chondrites (Browning* et al., *1996).*

Krot *et al.*, 1998). Aqueous alteration of Semarkona has converted the glassy mesostasis of a few chondrules to phyllosilicates, which sum to around 80%, and caused the deposition of calcite in the matrix (Hutchison *et al.*, 1987). The reaction of olivine with water to produce hydrated minerals has been experimentally investigated by Johannes (1968).

In addition to preaccretionary processes, aqueous alteration has played a major role in determining the oxygen isotope properties of carbonaceous chondrites and there is also evidence that thermal dehydration has affected some meteorites (Clayton and Mayeda, 1999).

Most authors have concluded that the aqueous alteration observed in meteorites occurred on the meteorite parent body, however there are some that argue that the reaction of water with the meteorite solids occurred in the nebula. Figure 3.5 shows typical petrographic evidence for aqueous alteration being nebular. In this case the authors suggested that the rim of alteration around a broken mineral grain suggested alteration in the nebula. Others used the same images to suggest an alternative sequence in which alteration occurred on the parent body. This ambiguity is typical of much petrographic interpretation and is a problem traditionally faced by petrologists, whether discussing terrestrial or extraterrestrial problems. For this reason it is important to separate description from interpretation.

3.2.3 *Long-lived radionuclides*

Unique clues to chondrite history are available through their chronolgy. U–Th, ^{147}Sm, ^{87}Sr, and ^{40}K are long-lived nuclides used in meteorite studies for radiometric dating. The basic principle is the same in all cases but the method of application varies considerably from system to system and some systems can be applied in many ways, for instance some U–Th methods rely on measurements of Pb isotopes produced and are referred to as Pb–Pb methods (Tatsumoto *et al.*, 1976; Tilton, 1988a; 1988b). Essentially, the amount of daughter isotope is measured in relation to the amount of the parent and this provides an indication of age, provided the initial amount of the daughter and the rates of decay are known.

3.2 Formation history

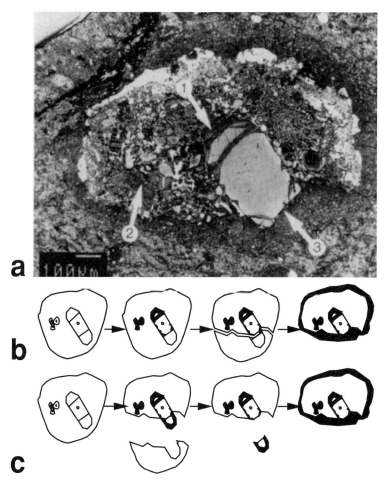

Figure 3.5 Summary of typical petrographic evidence for whether aqueous alteration occurred in the nebula or on the meteorite parent body. (a) A large chondrule in the Murchison CM chondrite shows aqueous alteration around the entire chondrule and around olivines in the chondrule (the altered material is marked 1). Small olivines in the chondrule are almost entirely altered (2), while a large olivine grain in the chondrule is aqueously altered on all but one side (numbered 3 in the figure). (b) Metzler *et al.* (1992) suggested that the olivines underwent aqueous alteration in the nebula prior to incorporation into the chondrule. (c) An alternative possibility is that the large olivine grain fragmented after aqueous alteration on the parent body.

Chondrites have been dated by the Sr, Nd, and Pb methods and generally have an age of 4.56 Ga, but the range is large and encompasses the range of ages shown by refractory inclusions and differentiated meteorites. St. Séverin has been especially thoroughly studied and yields an age of 4.552 ± 0.003 Ga (Tilton, 1988b). This is equal to the age of the Earth determined by isotopic dating and to astrophysical estimates for the age of the Sun.

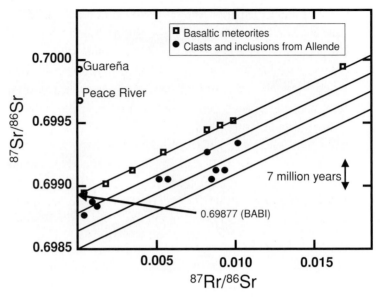

Figure 3.6 Rubidium–strontium evolution diagrams for samples with low levels of Rb showing differences in the initial ratios of $^{87}Sr/^{86}Sr$ (Gray and Papanastassiou, 1973). Since ^{87}Rb decays from the moment it is created, this ratio increases with time and provides a means of dating small time intervals during the formation of the meteorites. The acronym BABI (basaltic achondrites best initial) and the term basaltic meteorites refer to a class of igneous meteorites from the asteroid Vesta. The high $^{87}Sr/^{86}Sr$ ratio of the Guareña meteorite is thought to reflect metamorphism of this meteorite about 80 million years after the other materials formed.

Very small differences in formation time can be determined from the initial $^{87}Sr/^{86}Sr$ ratios. Thus it seems that the basaltic volcanism that produced the differentiated meteorites occurred after the formation of the refractory inclusions in Allende and that the relatively intense metamorphism experienced by the chondrites occurred ~80 million years after their formation (Fig. 3.6; Gray and Papanastassiou, 1973).

Additional information can be obtained from the decay of ^{40}K, especially if the ^{40}Ar–^{39}Ar method is used. In this case, instead of measuring parent ^{40}K directly, the sample is placed in a nuclear reactor where some of the ^{39}K (usually in simple ratio to ^{40}K) is converted to ^{39}Ar by an (n, p) reaction. Then the ^{39}Ar and ^{40}Ar can be determined in the same step-wise gas release experiment, where the complications of comparing different techniques are avoided and any unwanted components (like terrestrial contamination) can be isolated in the different temperature steps. When the gases released by several temperature steps yield the same age, it is possible to have confidence in that age.

A plot of K–Ar ages determined by the Ar–Ar method is shown in Fig. 2.6 and was discussed earlier. Two peaks are present, one at 4.6 Ga which is presumably the ages of the rocks – the time since the retention of ^{40}Ar began which is presumed to be the time at which the meteorites formed as the solid objects we see today – and a second peak at about 500 Ma. Most of the meteorites making up the younger peak are black in color, and contain numerous indications of a violent high-temperature event. These effects can be reproduced in the laboratory by subjecting normal meteorites to a sudden violent shock impulse such as projecting steel plates at them in a controlled fashion using chemical explosives. In fact the mineralogical and petrographic changes can be "calibrated" in terms of shock pressures quite precisely. Thus it is concluded that the 500 Ma peak represents a major collisional breakup of the L chondrite parent body 500 Ma ago (Urey, 1956; Heymann, 1967; Watanabe *et al.*, 1986; Bogard, 1994; 1995).

3.2.4 Short-lived radionuclides

Nuclides such as ^{26}Al, ^{53}Mn, ^{107}Pd, ^{129}I, ^{244}Pu, and ^{146}Sm were present when the meteorites formed but are now extinct; however their daughter isotopes are still present (Wasserburg, 1985). They are interesting for two reasons. First, they started to decay as soon as they were made so the fact that they were present when the meteorite formed means that there was a short interval between the end of element formation and meteorite formation. If it were possible to determine how much of the nuclide was originally produced, it would be possible to calculate this "formation interval." Several estimates have been made using various assumptions, but there is no agreement in ages obtained from different isotopes. Second, from the relative amounts of these isotopes present at the time of formation, determined from the daughters, it is possible to detect very small differences in the time of formation of different materials. For example, from the relative amounts of ^{129}Xe, the daughter isotope of ^{129}I, it has been concluded that the chondrites formed over a 20 Ma time interval and that the order of formation was C, H, L, LL, E chondrites, aubrites, and irons (Fig. 3.7). There have been many studies of I–Xe systematics of chondrites (e.g. Podosek, 1970; Podosek and Swindle, 1988a; 1988b; Swindle and Podosek, 1988).

Aluminum-26 is particularly interesting because it decays relatively quickly and because Al is a cosmically abundant element (Loveland *et al.*, 1969). It could therefore represent a significant heat source for the early Solar System. The important question is: how much ^{26}Al was there, relative to the major isotope ^{27}Al, when these materials accreted? ^{26}Al/^{27}Al ratios can be determined in high-Al – low-Mg materials (^{26}Mg being the daughter product of ^{26}Al) like the minerals in refractory inclusions, where it is found to be about 5×10^{-5} (MacPherson *et al.*, 1995). This

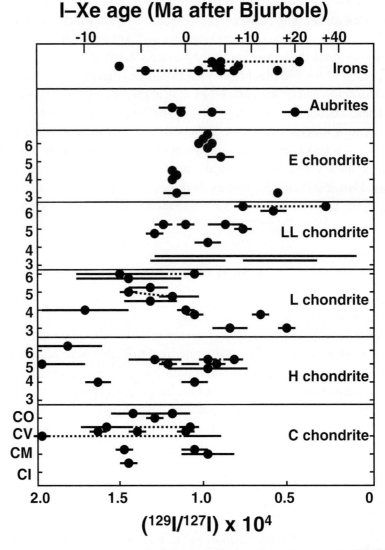

Figure 3.7 Compilations of $^{129}I/^{127}I$ ratios and apparent I–Xe ages for chondrites and other meteorites (after Swindle and Podosek, 1988). Time increases from left to right on the diagram. ^{129}I is now extinct and its original abundance in the meteorite is calculated from stable ^{129}Xe. Since ^{129}I decays with an 18 million year half life, the $^{129}I/^{127}I$ ratio reflects formation time. Thus, approximately speaking, the order of formation goes from bottom to top in this diagram.

is a high value, and if all asteroids contained this ratio of ^{26}Al to ^{27}Al when they formed every asteroid larger than 5 km would have been heated enough to make it melt (Lee et al., 1976). This is very unlikely, so it can be assumed that most materials formed later than the refractory inclusions, after much of the ^{26}Al had decayed away, and formation times for refractory inclusions obtained with long-lived nuclides seems to be consistent with this (Swindle et al., 1996). It has been suggested that the formation age of asteroids increases with distance from the Sun and that the amount of ^{26}Al in them at the time of formation decreased. Thus asteroids in the Inner Belt are more likely to have been heated than asteroids in the Outer Belt and this explains why the proportion of S asteroids to C asteroids decreases with distance from the Sun (Grimm and McSween, 1993).

3.2.5 Cosmogenic nuclides

Cosmogenic nuclides are those produced by nuclear reactions between energetic particles from outside the meteorite and the atoms from which the meteorite is composed. When properly understood in terms of laboratory experiments and theoretical modeling, they shed unique insight into many phases of meteorite history. In Section 2.1.4 we showed how cosmic ray exposure ages provide the best available evidence that most meteorites came from relatively few parent objects. The presence of cosmogenic nuclides is often the best evidence for some materials being extraterrestrial, for example interplanetary dust particles, or for not being extraterrestrial, for example tektites. Cosmogenic nuclides also enable the determination of preatmospheric shape, transit times from the Earth or Moon for meteorites originating on these bodies, and duration of exposure on the surface of a parent body, where a meteorite has been exposed to multiple episodes of cosmic radiation. They also enable insights into the activity of the Sun at earlier times. Related physical phenomena are charged particle tracks and thermoluminescence events that enable similar and sometimes complementary information on the history of these materials.

3.2.6 Presolar grains

A number of refractory grains have been discovered in meteorites in trace amounts that contain unusual isotopic properties. In general, these properties are not well-understood, although in some cases they seem to be consistent with certain phases of nucleosynthesis in a variety of environments associated with different kinds of star (Clayton et al., 1985; Lee, 1988). Trace amounts of interstellar grains in meteorites might not be surprising, especially in view of their refractory and almost indestructable nature, and such grains form a direct link between our own Sun and adjacent stars. Typically, the grains are graphite, diamond, silicon carbide, or alumina. The significance of these grains to chondrule and chondrite formation might be of more philosophical than practical significance, because these ideas remind us of the possibility of many chondrite and chondrule properties being inherited from the interstellar or at least pre-planetesimal stage of Solar System evolution.

3.2.7 Interclass relationships

It is probably useful to try to relate the classes of meteorites to each other; sometimes this seems fairly straightforward and at other times it seems speculative. The three classes of ordinary chondrite, H, L, and LL, have many properties in common and so are clumped together as ordinary chondrites. While there are differences, they are differences of degree and not kind (chondrule sizes, metal-to-silicate ratios, degree of oxidation or reduction) so it seems reasonable to link them. Similarly, the EH and EL chondrites are qualitatively similar, and are probably related. Linkage between the various carbonaceous chondrite classes is more problematic, and so is linkage between chondrites and other classes, say iron meteorites or stony iron meteorites. On the basis of trace element geochemistry, H chondrites have been linked to IIE iron meteorites. At one time IIIAB iron meteorites were linked with L chondrites by the similarity in L chondrite Ar–Ar ages and IIIAB iron meteorite cosmic ray exposure ages, but this similarity no longer seems to be true.

A point to stress is that although there may be no reason to link two classes, there is also no reason to assume they are from different parent bodies. Without knowing the causes of the differences in properties between two classes, for example degree of oxidation, oxygen isotope ratios, or metal-to-silicate ratio, it seems unwise to assume anything about interrelationships and linkages.

3.3 The challenge

In scientific research it is often said that the challenge is not only in answering questions, but also in asking the right questions. Having attempted to summarize the information we have on these primitive Solar System materials it seems to me at least arguable that the main questions are:

- What was the origin of the chondrules?
- What caused the metal–silicate fractionation?

I suspect that all of the other properties of chondrites and the individual chondrite classes are related to these two points. Concentrating on these questions will yield a certain view on the origin of chondrules and chondrites whose ultimate success will be measured by its ability to explain the facts and make predictions about future discoveries. In principle, analyzing other questions should ultimately yield the same answers, although the path might be different. But now let's look more closely at the chondrules, and explore their origin, before looking at theories for the metal–silicate fractionation and the challenge of bringing it all together.

4
Chondrules and their main properties

4.1 The diversity of chondrules

Chondrules are highly diverse in their properties. They all appear to have been melted to some degree at some point; some are clearly melt droplets that have crystallized; some appear to be igneous systems that have been subsequently rounded. Most are deficient in metal and sulfide in comparison with the host rock, some have additional compositional differences. Some have thick rims of material, some do not, and in some cases those rims are rich in metal and sulfide, and in some cases they are not. The nature of the chondrules, and their diversity, tells us something about conditions in the early Solar System. Chondrules are the major component of chondrites and must account for many of their bulk properties, as does their relative abundance in the various classes of meteorites (Huang *et al.*, 1996b). It also seems clear that accounting for the diversity of properties will provide insights into their formation process.

Attempting to understand objects as diverse as chondrules starts with their classification and there have been many proposed schemes. However, all the schemes have common features and it is possible to trace a development of ideas. In this way I hope to focus on the important observations without getting bogged down in the plethora of detail, much of which may be extraneous. There has been a long-term trend of textural classification schemes giving way to composition-based schemes as analytical instruments improved and trends became clearer. This long-term trend also reflects the application of the schemes to a greater variety of chondrules and host chondrites, particularly metamorphosed chondrites. Representative chondrule textures are shown in Figs. 4.1 to 4.3.

74 *Chondrules and their main properties*

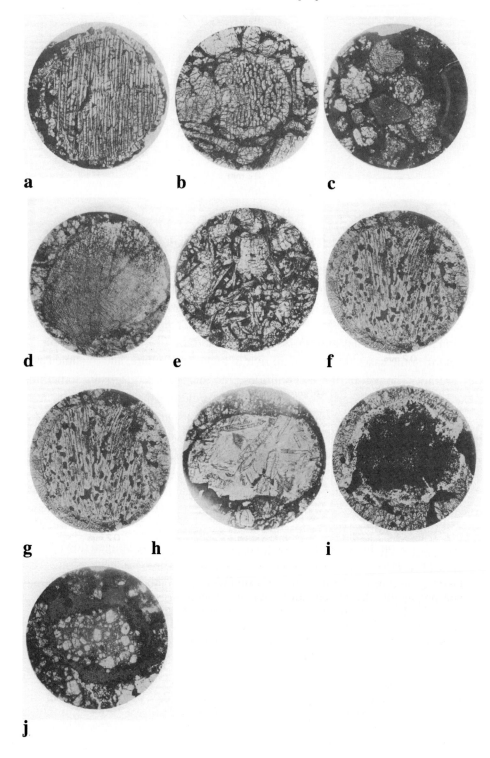

a

b

c

d

e

f

g

h

i

j

4.2 Chondrule groups

4.2.1 Tschermak scheme

Tschermak (1885) produced the first and still the best catalog of chondrule textures and in presenting the photomicrographs divided the chondrules into olivine chondrules (Fig. 4.1a, b), pyroxene chondrules (Fig. 4.1c, d), olivine–pyroxene chondrules (Fig. 4.1e), augite chondrules (Fig. 4.1f), plagioclase chondrules (Fig. 4.1g), glass chondrules (Fig. 4.1h), shock-blackened chondrules (Fig. 4.1i), and chondrules with iron-rich rims (Fig. 4.1j). Thus most of the major mineralogical and textural properties of chondrules were known before the end of the nineteenth century.

4.2.2 Merrill scheme

Merrill (1920) was probably the first person to divide chondrules into porphyritic (Fig. 4.2a–g) and non-porphyritic (Fig. 4.2h–1) forms. He also described rims, which he called "secondary borders" (Fig. 4.2j), chondrules with plastic deformation (Fig. 4.2k) and compound chondrules (Fig. 4.2l) one chondrule inside another.

4.2.3 Kieffer–King scheme

Kieffer (1975) and King (1983b) referred to droplet and lithic chondrules suggesting that while droplet chondrules were crystallized melts, lithic chondrules were abraded fragments of igneous rocks. To a reasonable approximation, these categories matched Merrill's non-porphyritic and porphyritic chondrules.

Figure 4.1 Representative chondrule textures according to Tschermak (1885). (a) Lamella olivine chondrule in Mezö Madaras (Fig. 38 in Tschermak, field of view 425 µm). (b) Chondrule in Dhurmsala consisting of a single olivine crystal in brown glass (Fig. 36, 2.4 mm). (c) Excentroradiating pyroxene chondrule (right) and porphyritic olivine chondrule with turbid mesostasis (left) in Homestead (Fig. 28, 1.3 mm). (d) Excentroradiating pyroxene chondrule in Dhurmsala (Fig. 53, 1.4 mm). (e) Olivine (wide crystals) – pyroxene (narrow crystals) chondrule in Knyahinya (Fig. 44, 1.2 mm). (f) Augite, sometimes enclosing olivine, in a glass in chondrules in Rennazo. Note also the opaque matrix surrounding the chondrules. Polarized light with crossed Nicols (Fig. 58, 1.2 mm). (g) Rare plagioclase-rich chondrule, with occasional olivine, and abundant sulfide in Dhurmsala (Fig. 64, 2.2 mm). (h) Devitrified glass chondrule in Rennazo or Lancé (Fig. 69, 1 mm). (i) Blackened chondrule with maskelynite in Château-Renard (Fig. 67, 540 µm). (j) A porphyritic olivine chondrule in Monroe with an iron shell (Fig. 75, 1.4 mm).

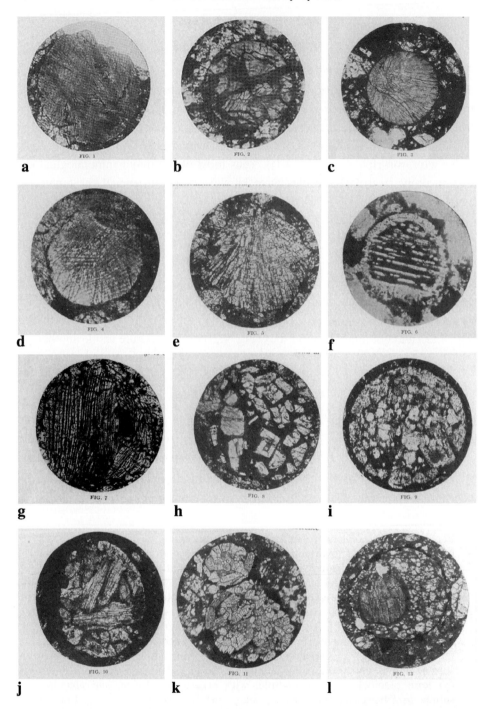

4.2.4 Dodd scheme

Dodd (1978a; 1978b; 1981) described "lithic" or "clastic" chondrules in his 1981 textbook, but in his descriptions of the chondrules in several type 3 ordinary chondrites he also divided them into metal-rich and metal-poor microporphyritic chondrules (Dodd, 1971; 1973).

4.2.5 Gooding–Keil scheme

Gooding and Keil (1981) divided the chondrules into porphyritic and non-porphyritic and then described the mineralogy and texture of the chondrule. Porphyritic chondrules are porphyritic olivine–pyroxene, porphyritic pyroxene, while non-porphyritic chondrules are cryptocrystalline, radiating pyroxene, barred olivine (Gooding and Keil, 1981). This scheme sees a return to Tschermak's use of mineralogy in chondrule classification.

4.2.6 Wood–McSween scheme

Following early work of Wood (1962) on olivine compositions, McSween (1978) and McSween *et al.* (1983) divided the chondrules in carbonaceous chondrites into types I (Fe-poor olivines), II (Fe-rich olivines) and III (non-porphyritic) chondrules. This represents the first use of mineral composition in chondrule classification.

4.2.7 Scott–Taylor–Jones schemes

These authors extended the Wood–McSween scheme to chondrules in other chondrite classes and they made the scheme considerably more elaborate by including

---

Figure 4.2 Representative chondrule textures according to Merrill (1920). (a)–(g) Glassy, crypotocrystalline, radiating and barred (i.e. non-porphyritic) forms. (h)–(j) Porphyritic forms. (k, l) Compound chondrules. Merrill did not give scale bars, but it seems probable that each field of view is about 0.5 mm in diameter. In all cases the Nicols are crossed, so that olivine and pyroxene appear gray and glass is dark. (a) Cryptocrystalline chondrule in Barratta (Fig. 1 in Merrill). (b) Cryptocrystalline chondrule in Cullison, Kansas (Fig. 2). (c) Radiating chondrule in Elm Creek (Fig. 3). (d) Radiating chondrule in Hessle (Fig. 4). (e) Radiating chondrule in Parnallee (Fig. 5). (f) Barred olivine chondrule in Beaver Creek (Fig. 6). (g) Barred olivine chondrule in Hendersonville (Fig. 7). (h) Porphyritic chondrule in Tennasilm (Fig. 8). (i) Holocrystalline polysomatic chondrule in Barratta (Fig. 9). (j) Holocrystalline polysomatic chondrule in Elm Creek (Fig. 10). (k) Compound chondrule showing plastic deformation in Parnallee (Fig. 11). (l) Compound chondrule in an unidentified stone (Fig. 13).

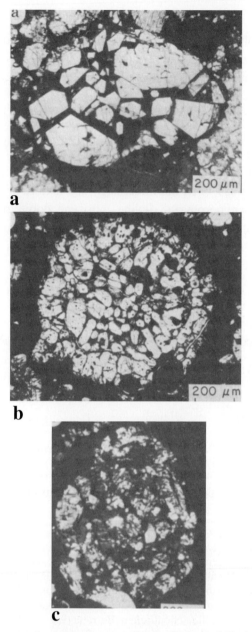

Figure 4.3 Representative textures according to Scott and Taylor (1983). (a) A high-FeO porphyritic chondrule in Semarkona LL3.0 chondrite. (b) A low-FeO chondrule in Semarkona. (c) A poikylitic pyroxene chondrule in Semarkona. All images are to the same scale, where the horizontal field of view of the top image is about 1.0 mm.

Table 4.1 *Definition of the chondrule groups and frequency of occurrence in two primitive ordinary chondrites, Krymka (LL3.1) and Semarkona (LL3.0).*[a]

	Mesostases		Olivine		Frequency[c]	
CL	Composition[b]	CL	% FeO	% CaO	Krymka (51)	Semarkona (76)
A1 yellow	Pl(An>50%)	red	<2	>0.17	3.6	10.5
A2 yellow	Pl (An>50%)	none/dull red	2–4	0.1–0.2	0.0	25.0
A3 blue	Pl (An>50%)	red	<4	>0.2	33.3	0.0
A4 blue	Pl (An>50%)	none/dull red	>4	0.16–0.3	7.3	0.0
A5 blue	Pl (An<50%)	none	>4	<0.25	14.5	5.0
B1 none	>30% Qtz	none	4–25	0.08–0.3	0.0	56.9
B2 none	30–50% Qtz	none/dull red	10–25	0.08–0.3	36.4	0.0
B3 purple	15–30% Qtz	none	15–20	<0.08	0.0	2.6

[a] Phase compositions given as a guide, the group fields are not rectangular (see Sears *et al.*, 1992; 1995b for details). "CL" refers to cathodoluminescence color.
[b] Normative composition (wt %) of the mesostasis: Pl, plagioclase; An, anorthite; Qtz, quartz.
[c] The figures in parentheses indicate the number of chondrules on which the statistics are based. Semarkona data from DeHart *et al.* (1992); Krymka data from Huang *et al.* (1993b).

textural properties (Scott and Taylor, 1983; Jones and Scott, 1989; Jones, 1990; 1994). Thus they have type I (Fe-poor) and type II (Fe-rich), subdivided into A (olivine rich), AB (poikylitic pyroxene), and B (pyroxene-rich) based on mineralogy and texture. Thus the scheme is essentially textural and mineralogical, with fine-grained olivine chondrules being type I, coarse-grained olivine-rich being type II, and a third type being pyroxene-rich (Fig. 4.3). However, these definitions really only apply satisfactorily to unmetamorphosed meteorites because while chondrule composition changes with metamorphism, textures are essentially unchanged until the highest intensities of metamorphism are reached.

4.2.8 Sears et al. scheme

Sears *et al.* (1992) broke away from texture to rely entirely on mineral compositions for the classification of chondrules. Chondrules are divided into groups depending on the composition of their two major phases – olivine (or pyroxene) and the mesostasis (groundmass), both of which are genetically important (Sears *et al.*, 1992; 1995b; Table 4.1). Compositional fields on plots of the normative quartz–albite–anorthite ternaries for mesostasis and CaO vs FeO for olivine (Fig. 4.4)

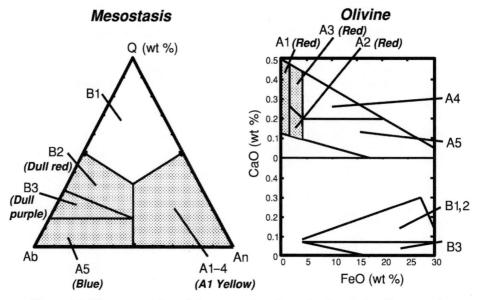

Figure 4.4 The composition of the two major phases in chondrules (Sears et al., 1992). Albite–anorthite–quartz ternaries for the mesostasis and calcium against iron in the olivine of chondrules in ordinary chondrites. Their cathodoluminescence colors are given, the fields not labeled with colors do not display any cathodoluminescence. The composition of the phases gives them unique appearance under the cathodoluminescence microscope so that most classes of chondrule can be readily identified.

enable the identification of eight discrete classes of chondrule. Four of the classes (A1, A2, A5, and B1) are found in primitive i.e. essentially unmetamorphosed (Lu et al., 1990; 1992) chondrites and four (A3, A4, B2, and B3) are found in metamorphosed chondrites and are the result of metamorphism of the primitive groups. Group A5 appears in both metamorphosed and unmetamorphosed chondrites although the degree of compositional heterogeneity decreases with metamorphism. The compositional classification scheme for chondrules is compared with previous schemes in Table 4.2. The cathodoluminescence of chondrules and relative abundance of chondrule classes is so sensitive to metamorphism that the scheme can be used to assign petrographic types to type 3 ordinary chondrites with a precision comparable to that of induced thermoluminescence (Sears et al., 1995b).

4.3 Composition of chondrules

4.3.1 Bulk compositions of chondrules

The bulk compositions of chondrules have been reviewed by Grossman and Wasson (1983b), Kallemeyn (1988), and Huang et al. (1996b). Most of the data in the literature have been obtained by instrumental neutron activation analysis (INAA),

Table 4.2 References reporting analyses of chondrules.

Technique[a]	Class					
	Enstatite chondrites		Ordinary chondrites		Carbonaceous chondrites	
	Reference	Samples	Reference	Samples	Reference	Samples
INAA	Grossman et al. (1985)	Qingzhen	Evensen et al. (1979)	Richardton	Rubin and Wasson (1986)	Murray
			Gooding et al. (1980)	UOC	Rubin and Wasson (1987a)	Ornans
			Grossman et al. (1979)	Chainpur	Rubin and Wasson (1987b)	Allende
			Grossman and Wasson (1982)	Chainpur	Simon and Haggerty (1980)	Allende
			Grossman and Wasson (1983a)	Semarkona		
			Huang et al. (1993b)	A3 chondrules		
			Huang et al. (1994b)	A5 chondrules		
			Lu (1992); Huang et al. (1996b)	Semarkona and Krymka		
			Hamilton et al. (1979)	Parnallee		
			Fujimaki et al. (1981)	ALHA77015		
DEMPA	Ikeda (1983; 1989)	Enstatite chondrites	Ikeda (1983)	Ordinary chondrites	McSween (1977d)	Carbonaceous chondrites
			Wlotzka (1983)	Ordinary chondrites		
IDA					Misawa and Nakamura (1988)	CO chondrites
					Nakamura and Matsuda (1989)	Allende

[a] INAA, instrumental neutron activation analysis: DEMPA, defocussed electron microprobe analysis: IDA, isotope dilution analysis.

Table 4.3 *Comparison of the compositional classification scheme for chondrules with previously proposed schemes.*

Compositional class	Approximate previous equivalents
A1	Includes some of the droplet chondrules of Kieffer (1975), some of the non-porphyritic pyroxene chondrules of Gooding and Keil (1981), the type I chondrules of McSween (1977b), metal-rich microporphyritic chondrules of Dodd (1978a; 1978b), and the type IA chondrules of Scott and Taylor (1983).
A2	Includes the poikylitic pyroxene and type IB chondrules of Scott and Taylor (1983) and many of the type IAB chondrules of Jones (1992; 1994).
A5	There appear to be no previous observations of this chondrule group in unmetamorphosed meteorites.
B1	Dodd's (1981) "lithic" or "clastic" chondrules and Dodd's (1978a; 1978b) metal-poor microporphyritic chondrules are included in this group, as are the type II chondrules of McSween (1977c), Scott and Taylor (1983), and Jones (1990).

but defocussed electron microprobe and isotope dilution mass spectrometry methods are also important (Table 4.3).

The compositional trends are shown in Fig. 4.5. The reduced, or low-FeO, group A1 chondrules from all chondrite classes show major depletions in volatile lithophile elements Cr, Mn, Na, and K. In contrast, the oxidized, high-FeO, group B1 chondrules have CI proportions in all these elements. Chondrules of all classes show major depletions in siderophile and chalcophile elements, especially the volatile elements Ga, Sb, Se, and Zn. Group A5 chondrules (not shown in Fig. 4.5) have bulk compositional properties resembling group B1 chondrules.

A convenient means of looking for correlations between large data sets is by performing factor analysis. This technique clumps together elements or other analytical data that show strong correlations, and thus helps identify the processes likely to have affected the samples. Grossman and Wasson (1983a) performed factor analysis on their analytical data for the Semarkona chondrules and found that four factors could explain the trends in the data: the metal/sulfide abundance; the abundance of refractory lithophiles; the abundance of volatile lithophiles; and the ratio of olivine to pyroxene (Fig. 4.6). The question is, what caused the variations in these factors?

4.3.2 Laboratory experiments and chondrule compositions

While there has been considerable discussion of evaporative loss of elements during chondrule formation based on the meteorite data (e.g. Dodd, 1978a; 1978b; Hewins,

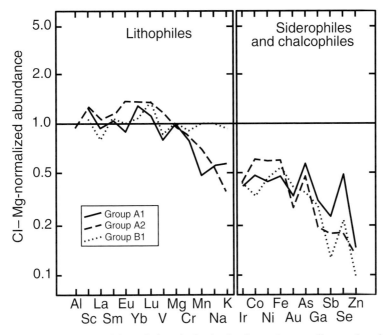

Figure 4.5 Bulk composition of chondrules in the Semarkona ordinary chondrite, sorted by the three compositional classes A1, A2, and B1 (Huang et al., 1996a). The fourth group found in unaltered meteorites, group A5, has compositions similar to group B1. For group B1 (and A5), lithophile elements are present in the same proportions in chondrules as they are in CI chondrites (after correcting for volatile element differences by normalizing to Mg), whereas for groups A1 and A2 abundance decreases with increasing volatility. All chondrule groups show major depletions in siderophile and chalcophile elements compared to CI chondrites, and again the depletion increases with increasing volatility (from Sears et al., 1996a).

1991), insights into deciphering the elemental abundance patterns in chondrules, and in understanding chondrule behavior during formation, are best obtained by thermodymanic calculation and by laboratory experimentation. Thermodynamics assumes that the various phases are at equilibrium, a reasonable first approximation, while laboratory experiments can quantify the thermodynamics and even explore some non-equilibrium effects. Most work has been performed in small laboratory furnaces, but King (1982) used a large solar furnace and was able to cause 90% evaporative loss and produce refractory residues. FeO-poor chondrule analogs could be produced from initially Fe-rich materials (King, 1983b).

Smaller laboratory experiments have been reported by Gooding and Muenow (1976; 1977), Donaldson (1979), Hashimoto et al. (1979), Hewins et al. (1981), Tsuchiyama and Nagahara (1981), Tsuchiyama et al. (1981), Hashimoto (1983), Nagahara (1986), Myson and Kushiro (1988), Nagahara and Kushiro (1989), Nagahara et al. (1989a; 1989b), Shimaoka and Nakamura (1989), and Tissandier

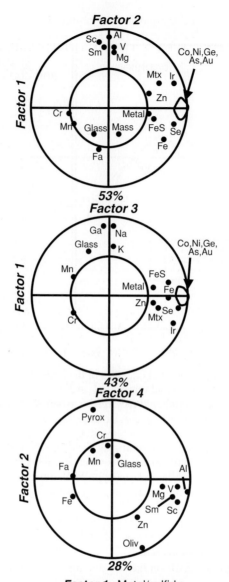

Figure 4.6 The results of factor analysis of Semarkona chondrule compositions (Grossman and Wasson, 1983a). Factor analysis groups elements according to the degree to which they correlate among the chondrules analyzed, thus siderophiles tend to correlate closely because they remain together in the metal phases throughout chondrule history. As might be expected, elements in the metal and sulfide remain together, as do volatile lithophiles and refractory lithophiles, but these three components vary independently of each other. Additionally, olivine and pyroxene ratios vary independently. Whether these components were present before chondrule formation, or were produced during chondrule formation, is not clear from these data.

4.3 Composition of chondrules

et al. (2002). Hashimoto *et al.* (1989) and Davis *et al.* (1990) have made measurements of isotopic fractionation during evaporation and Yu *et al.* (1995) have studied oxygen isotope exchange. The results of such evaporation experiments are summarized in Fig. 4.7. These laboratory measurements show that while refractories Al and Ca are not appreciably lost by evaporation during chondrule formation, Mg is slightly lost, Si is lost a little more readily than Mg, and Fe shows major evaporative loss – usually by reduction of FeO to Fe, Fe being a fairly volatile element. Sodium and elements of similar volatility should have been even more readily lost during chondrule formation, the amount of loss increasing as oxygen pressure decreases or temperature increases. Major increases in oxidation conditions (i.e. increases in $P(O_2)$) are required to prevent Na loss during chondrule formation. Similarly, Na and Si readily reenter the chondrule in laboratory experiments if present in the gas in even moderate concentrations (Lewis *et al.*, 1993; Georges *et al.*, 2000; Tissandier *et al.*, 2002). There is petrographic evidence for alkalis and chlorine entering the chondrules during crystallization (Bridges *et al.*, 1997).

The laboratory experiments also confirm thermodynamic predictions that after loss of alkali metals, iron should be reduced and Fe_m evaporation occurs that would increase the olivine–pyroxene ratio, and then silicon would be lost and this would partially restore the olivine–pyroxene ratio (Fig. 4.7b).

So, how do we heat chondrules enough to cause the droplet shapes and the melt properties and to change compositions in one case and not the other? The situation is summarized in the form of phase diagrams in Fig. 4.8 (Kimura and Yagi, 1980; Nagahara *et al.*, 1994). In the case of the reduced, FeO-poor, group A chondrules there was partial evaporation, while the FeO-rich, non-refractory group B were not heated to the same degree. But we are still faced with retaining the Na in these group B chondrules. Many authors argue that $P(O_2)$ was increased many orders of magnitude (Fig. 3.1a, b), perhaps the event occurred in a dust-enriched environment, while others have argued for supersaturation as a means of suppressing Fe formation and increasing the formation of FeO (Blander and Katz, 1967; Fig. 3.1c). Petrographic evidence for supersaturation and supercooling of FeO-rich chondrules has been discussed by several authors (Sorby, 1877; Lofgren and Russell, 1986; Jones, 1990).

Using the compositional classification scheme it is concluded that while the volatile-poor classes, A1 and A2, were formed by reduction of FeO and major evaporative loss from precursors originally resembling those of CI chondrites, evaporative loss from group B1 chondrules was restricted to only the most highly volatile trace elements like Ga, Sb, Se, and Zn. Thus the process that formed chondrules was capable of acting with a variety of intensities. Since chondrules are the major structural component in chondrites, it seems at least arguable that reduced state variations in chondrules are responsible in some part for much of the redox differences between the chondrite classes.

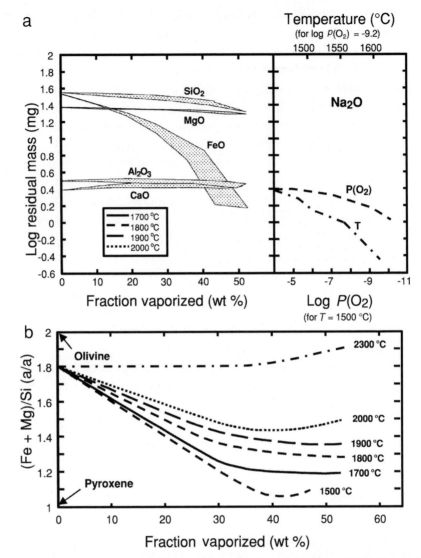

Figure 4.7 The results of laboratory heating of mixtures with approximately chondrule compositions (Huang et al., 1996a). (a) Volatiles are lost: first the alkali metals, then Fe (after reduction), then Si, and then Mg. Ninety percent mass loss results in residues similar to the refractory inclusions (Marvin et al., 1970). (b) Loss of Fe and then Si causes the olivine–pyroxene ratio to decrease and then rise again. Such experiments suggest that evaporation and reduction during chondrule formation can explain most of the compositional properties of chondrules, including a complex relationship between reduction and olivine–pyroxene ratio, the only problem being that two-thirds of the chondrules in ordinary chondrites retained significant amounts of sulfur and sodium.

4.3 Composition of chondrules

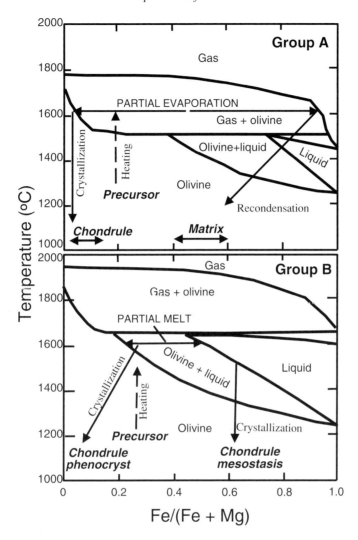

Figure 4.8 Proposed phase diagrams for chondrule and matrix formation according to Nagahara *et al.* (1994). For group A chondrules (FeO-poor) and matrix the ranges of Fe/(Fe+Mg) are shown. The precursor material with an Fe/(Fe+Mg) value of 0.2 is heated to ∼1600 °C and partially evaporates to produce a ferrous gas and magnesian residue that crystallizes to produce Fe-poor chondrule olivines. The gas recondenses to produce a matrix enriched in refractory Mg relative to the gas. For group B chondrules (Fe-rich) the same precursor material heated to ∼1600 °C but under higher pressure underwent melting without significant evaporation. Upon cooling it forms chondrules and mesostasis according to the olivine–liquid phase boundaries. A similar difference in final result could result from the precursor of group A and group B chondrules being heated at the same pressure but to different temperatures, i.e. being heated into the olivine–gas field to produce group A and into the olivine–liquid field to produce group B.

4.3.3 Chondrules as open or closed systems

There are two possibilities to explain the variations in compositions – either the range of compositional variation reflects the range of composition in the dust that was used to make the chondrules, or the range of compositions reflects differences in composition caused by the chondrule forming process. In other words, whether chondrules acted as open or closed systems during formation (Larimer and Anders, 1967; Wai and Wasson, 1977). "Open system" means that relatively volatile elements (like Fe, Na, and K) were lost and the chondrules underwent chemical reactions with species in the environment during formation, while "closed system" means that the various properties of the chondrules were inherited entirely from the precursors (Fig. 4.9). Of course, both processes probably occurred, so the discussion really concerns the relative importance of the two processes.

There are also a great many potential processes that might have occurred during the transient heating and cooling that was clearly the central feature of chondrule compositional variations. Such processes are reduction by ambient gases and carbon, impact by dust, recondensation of volatiles as surface rims, and diffusion of volatiles into the interiors of the chondrules (Hewins and Connolly, 1996; Yu et al., 1996). Most of these effects involve gas phases, but some will also involve liquid phases (Grossman, 1996b).

The question of whether chondrules behaved as open systems during formation or inherited their present compositions from the precursors was keenly debated in the early 1970s (Dodd and Walter, 1972; Walter and Dodd, 1972; Hewins, 1991). Compositional gradients in the contiguous mesostasis in which volatiles are enriched in the outer zone and refractories are enriched in the central zone, must indicate migration of the volatiles into the chondrule. Grossman (1996b) suggested that the volatiles were carried as saline solutions, but lack of aqueous alteration of these chondrules and the absence of such zoning in adjacent chondrules suggests that process occurred prior to incorporation and that the volatiles were the gas phase. This does not mean that the process occurred in the nebula prior to accretion. Several authors have described evidence for multiple recycling of chondrules (Alexander, 1994; 1996; Rubin and Krot, 1996), so that it is perfectly possible that gases coexisted with chondrules long after the initial formation of the chondrules.

The refractory composition, low-FeO silicates, relatively high metal abundance and low-Ni content of the metal of group A1, 2 chondrules are consistent with open-system behavior. A closed-system scenario would require that chondrule precursors were the products of earlier volatility–oxidation processes. The trend in the olivine–pyroxene ratio (which decreases from group B1 to A2 and then increases again in group A1), the smaller mean size of group A1 and A2 chondrules compared to group B1 chondrules, the relationships between oxygen isotope composition and

4.3 Composition of chondrules

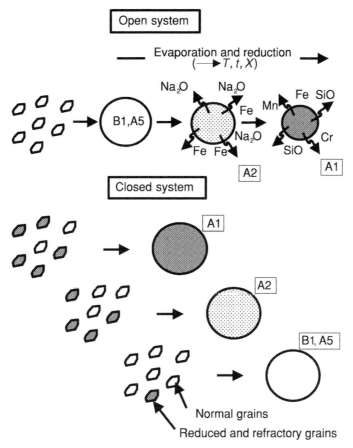

Figure 4.9 Schematic diagrams illustrating two options for the behavior of chondrules during formation. They either behaved as closed systems, in which case the great variety of chondrule properties were inherited from the precursors, or they behaved as open systems in which case many chondrite properties that are a reflection of chondrule properties are a result of chondrule formation processes.

chondrule size and peak temperature, diffusion of Na into chondrules, and the greater abundance of thick fine-grained rims around group A chondrules relative to group B chondrules are consistent with major evaporative loss, first of FeO and later SiO_2, accompanying the formation of groups A1 and A2 (Matsunami *et al.*, 1993; Huang *et al.*, 1996a). These properties are difficult or impossible to understand in terms of closed-system behavior, although Grossman (1996b) still makes the case. The current situation is summarized by Sears *et al.* (1996a) whose "truth table" is shown in Table 4.4. The most telling evidence is the correlation between chondrule mineralogy and redox state that could not arise from precursor mixing, and the compositionally zoned mesostasis of a few group A1 chondrules (Ikeda and Kimura, 1985; Matsunami *et al.*, 1993).

Table 4.4 *Truth table for chondrules being open or closed systems during formation (Sears et al., 1996a).*

Major chondrule properties	Open	Closed
Range of $P(O_2)$	Yes	Yes
$P(O_2)$-related patterns in mineralogy	Yes	No
Elemental abundance trends	Yes	Possibly
Differences in size	Yes	Doubtful
O-isotope trends	Yes	No
Recondensation of volatiles	Yes	No

4.4 Physical processes affecting chondrule history

4.4.1 Laboratory experiments and the textures of chondrules

Many laboratory experiments that reproduce the textures observed in chondrules (Hewins, 1988) have been performed. The data indicate a major role for nucleation centers, and they constrain temperatures and cooling rates of chondrules during formation (Hewins, 1988; Lofgren, 1989). Figure 4.10 is an attempt to summarize some of this work.

Laboratory experiments have also proved useful in understanding chondrule textures and factors that might affect them (Hewins, 1983; 1988; Connolly and Hewins, 1996). Nelson *et al.* (1972) produced artificial chondrules with chondrule

Figure 4.10 (a–d) Making chondrules with a CO_2 laser (From Keil *et al.*, 1973; see also Nelson *et al.*, 1972). (a) Spray of spherules formed by feeding a horizontal alumina rod into the focus of a carbon dioxide laser. Note the abrupt brightening of each trace as the spherules froze from the supercooled state. (b) 560 μm diameter spherule and (c) a 460 μm diameter spherule made from compressed forsterite–albite melted on an iridium wire in air and shaken off. (d) 420 μm spherule from an alumina drop shaken from a 0.010" sapphire rod and frozen in free fall in air. Keil *et al.* (1973) likened these objects (and many more) to the crystal-bearing spherules found in lunar breccias. (e–h) Textures of synthetic chondrules formed from olivine normative starting compositions after: (e) heating to the melting point and cooling at 10 °C/h; (f) heating to just below the melting point and cooling at 10 °C/h; (g) heating to just below the melting point and cooling at 100 °C/h; (h) heating to just below the melting point and cooling at 1000 °C/h (Hewins, 1988). (i) Textures of synthetic chondrules formed from pyroxene normative starting compositions after heating to a variety of starting temperatures (vertical columns) and cooled at 5, 100, and 2500 °C/h (horizontal rows) (Lofgren and Russell, 1986). Virtually all the textures observed in chondrules have been reproduced in the laboratory over a wide range of starting temperatures, cooling rates, and compositions. The most important factor seems to be the starting temperature as this determines the number of seed crystals (nucleation sites) for crystallization.

4.4 Physical processes affecting chondrule history

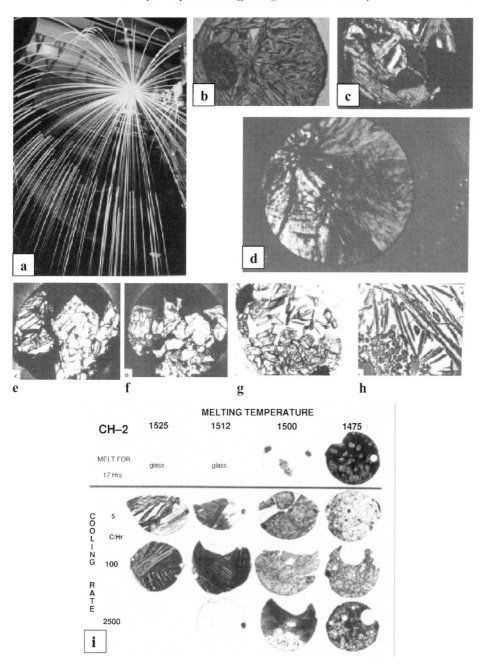

textures using a CO_2 laser (Fig. 4.10), while the textures of olivine-rich chondrules have been reproduced by Radomsky et al. (1986), Radomsky and Hewins (1987; 1988; 1990), Connolly et al. (1988; 1993), Lofgren (1989; 1996), and Jones and Lofgren (1993). Lofgren and Russell (1986) have investigated pyroxene-rich chondrules. Greenwood and Hess (1996) have investigated the kinetics and melting relationships of chondrules. Such experiments demonstrate the importance of nucleation centers (and thus peak temperature) in determining chondrule textures, and that bulk composition and reducing agents influence the final product. The search for analogous structures in synthetic chondrules to those observed in natural chondrules has been extended down to the nanometer scale (Weinbruch et al., 2001).

4.4.2 Temperatures, cooling rates, crystallization history, and relic grains

Temperatures for melts of chondrule compositions can be estimated from bulk compositions by the method of Herzberg (1979), from the temperatures required to produce certain textures in the laboratory experiments (Wood, 1979; Tsuchiyama and Nagahara, 1981; Fujii and Miyamoto, 1983; Hewins and Radomsky, 1990; Ruzicka, 1990; Hewins and Connolly, 1996; Weinbruch et al., 1998; Ferraris et al., 2002) and from the application of mineralogical thermometers (Saxena, 1976). For example, the distribution of Mg^{2+} between olivine and adjacent melt, K, is related to melting temperature, T, by:

$$\ln K = -\frac{14720}{T} + 6.806 \qquad (8)$$

Cooling rates can be determined from experimental reproduction of observed textures. The results are summarized in Table 4.5.

The unusual high Ca olivine in FeO-poor, refractory, group A chondrules has been the subject of many studies because it is not predicted by known phase relationships (Dodd, 1973). In fact, despite considerable effort, crystal–liquid equilibria in chondrules in general are not well-understood and do not follow expectations (Roedder and Emslie, 1970; Irving, 1978; Ford et al., 1983; Cirlin et al., 1985; Colson et al., 1986; 1988; Taylor and Cirlin, 1986; Dunn, 1987; Jurewicz and Watson, 1988). The problem is the lack of equilibrium and the unusual composition of the mesostasis. Some chondrules exist in which silica is present in the mesostasis (Olsen, 1983; Brigham et al., 1986). Refractory chondrules containing calcic plagioclase are fairly common (e.g. Grossman and Rubin, 1986).

Many non-refractory, high-FeO, group B chondrules contain "relic grains," that is grains that are clearly not related to the host chondrule and are a left-over of the original grains from which the chondrule was made (Nagahara, 1981; 1983a;

Table 4.5 *Temperatures and cooling rates for chondrules.*

Type of chondrule	Cooling rate (°C/h)	Ref.[a]	Temperature (°C)	Ref.[a]
Type I, FeO-poor, group A				
Porphyritic	100	1	1583±2	1
Barred	1000	1	1583±2	1
Porphyritic	24000–1200	2		
Barred	7200–300	2		
Radial	24000–1200	2		
Porphyritic	100	3	≤1590	3
Barred	1000–3000	3	>1590	3
Non-porphyritic			~1700	5
Type II, FeO-rich, group B				
Porphyritic	5	4		
Radial	5	4		
Porphyritic			~1300	5
Non-porphyritic			~1500	5
Unspecified	50–3000	6		
	5–10	6		

[a] Refs: 1, Radomsky and Hewins (1987); 2, Tsuchiyama *et al.* (1967); 3, Lofgren (1989); 4, Lofgren and Russell (1986); 5, Radomsky and Hewins (1988); 6, Ferraris *et al.* (2002). Faster cooling rate is for 1250–1200 °C and the slower rate is for 1000–650 °C.

1983b; Rambaldi, 1981; Kracher *et al.*, 1984; Watanabe *et al.*, 1984; Steele, 1985; 1986; 1988; Jones, 1996; Weisberg and Prinz, 1996; Jones and Danielson, 1997). The presence of relic grains in group B chondrules indicates that they were not completely melted during their formation. This is also additional evidence that chondrules undergo recycling during their formation.

4.4.3 Metamorphism and aqueous alteration of chondrules

The metamorphism of chondrules has been examined by many authors (Lux *et al.*, 1980; 1981; Sears *et al.*, 1984; 1995d; McCoy *et al.*, 1991; DeHart *et al.*, 1992). Olivine loses Ca and gains Fe during metamorphism and the mesostasis compositions move towards the normative albite corner of the albite–anorthite–quartz ternary. Thus, metamorphosed equivalents of the four primary chondrule groups (A1, A2, A5, and B1) are A3, A4, B2, and B3 with all chondrules finally assuming A5 properties, albeit more compositionally homogeneous than the A5 chondrules found in unequilibrated chondrites. The proportion of chondrules of various classes is highly dependent on the metamorphic intensity experienced by the host meteorite and can be used to assign a highly reliable petrographic type.

One feature that has attracted some interest is the presence of fayalite borders on the otherwise Fe-free olivine grains in chondrules. These clearly reflect a late-stage oxidation of the olivines, but how and when this occurred is unclear (Kring, 1987; Kring and Wood, 1987; Peck and Wood, 1987; Hua *et al.*, 1988).

When Hutchison *et al.* (1987) first described traces of aqueous alteration in unequilibrated ordinary chondrites, they mentioned several chondrules that had been altered. The mesostasis of the chondrules had been entirely converted to clay minerals and the electron microprobe analyses summed to ~80%, about 20% being undetected water. Subsequently the recent I–Xe ages of Semarkona chondrules, the "bleaching" of the outermost regions of chondrules (loss of mesostasis, alkalis, and Al) (Grossman, 1996a; Grossman *et al.*, 2000), and even the concentric compositional and physical zoning of chondrule mesostasis have been attributed to aqueous alteration, although the petrographic data are seldom as convincing as in the original Hutchison *et al.* (1987) paper.

4.5 Chondrule rims and matrix – implications for formation history

4.5.1 General comments

Merrill (1920) noted that many chondrules had secondary borders, now referred to as "rims," that resembled the matrix but sometimes included sulfide and metal. Tschermak (1883, Fig. 4.1) showed matrix in Tieschitz. Many authors have studied both the fine-grained rims around chondrules and the interchondrule fine-grained matrix and most have remarked on their similarity (King and King, 1981; Matsunami, 1984; Rubin, 1984b; Grossman, 1985; Alexander *et al.*, 1989; Rubin *et al.*, 1990; Brearley and Geiger, 1991; Metzler *et al.*, 1992; Sears *et al.*, 1993; Brearley, 1996). However, as Merrill and others have noted, some of the rims contain one or many zones of metal and sulfide; chunky rims of metal and sulfide have also been observed, such chondrules sometimes being referred to as "armoured." The metal and sulfide in chondrules has also been studied by several authors (Sears and Axon, 1975; 1976; Afiattalab and Wasson, 1980; Rambaldi and Wasson, 1981; 1982; 1984; Grossman and Wasson, 1985; Huang *et al.*, 1993a). The existence of small chondrules ("microchondrules," Rubin *et al.*, 1982; Christophe-Michel-Lévy, 1987) in the rims indicates that chondrule formation and rim formation were contemporary (Krot and Rubin, 1996).

Podolak *et al.* (1993) likened certain opaque rims on UOC (unequilibrated ordinary chondrites, another name for type 3 ordinary chondrites) chondrules to meteorite fusion crusts, since both have metal- and sulfide-rich zones. They suggested that opaque chondrule rims could have formed when the chondrules entered the atmosphere of their parent body (Fig. 4.11). Several authors have argued that

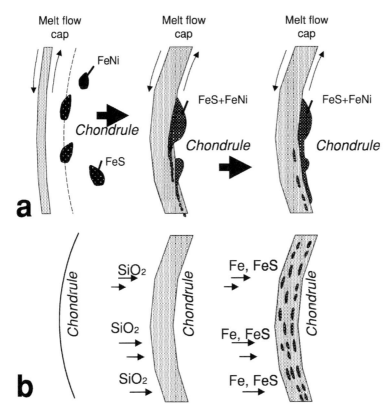

Figure 4.11 (a) Formation of opaque-rich rims on chondrules as a tumbling meteorite passes through a planetary atmosphere (Podolak et al., 1993). (b) Another possibility is that sulfur-rich vapors condensed onto the outside of the chondrule as a very fine-grained mixture of sulfides, metal, and silicates. Sometimes these rims were mildly heated (perhaps due to the residual heat in the chondrule) to cause coalescence of the sulfide and metal grains.

accreting parent bodies should have atmospheres (Ahrens et al., 1989). However, the body would have to be large (1000 km) and entry velocities of 2–4 km/s for chondrules and ~1 km/s for CAI are required.

Several authors have pointed out that the thickness of fine-grained rims around chondrules probably carries important information about their history. As an example, Morfill et al. (1998) has used the rim thickness to model solar nebula accretion processes; however his ideas neglect certain key properties of the rims. Figure 4.12 summarizes some of the information on these rims. The larger the chondrule, the larger the rim, but the ratio of rim thickness to chondrule size depends on the nature of the chondrule. Refractory volatile-poor reduced chondrules (group A) have thicker rims than volatile-rich oxidized chondrules (group B), suggesting that the rim is composed of elements that were evaporated from the chondrule and that

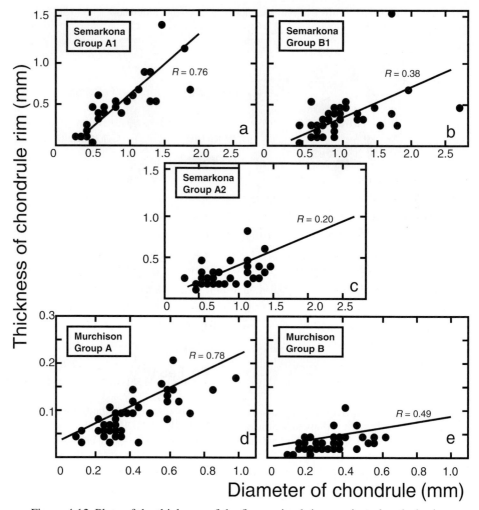

Figure 4.12 Plots of the thickness of the fine-grained rims against chondrule size for group A and group B chondrules in the Semarkona type 3.0 ordinary chondrite and the Murchison CM2 chondrite (Sears et al., 1993; Huang et al., 1996b). Several CM chondrites produce slopes very similar to those of Murchison. Group B chondrules invariably have rims that are thinner than group A chondrules of similar radius, i.e. they have shallower slopes on these plots.

subsequently condensed on the exterior of the chondrule. To a reasonable approximation, the rim plus chondrule composition of reduced chondrules is similar to the composition of oxidized chondrules, consistent with the idea that the rims are condensed chondrule volatiles. Morfill's ideas are difficult to reconcile with the dependence of rim thickness on chondrule type.

There has been considerable interest in fairly large "isolated olivine grains" in the matrix of several classes of chondrites, some authors suggesting that they are primary condensates from the primordial solar nebula and others arguing that they are grains that were formed in chondrules (Olsen and Grossman, 1978; Steele, 1986; Jones, 1996).

4.5.2 Ordinary chondrites

The nature of the matrix in ordinary chondrites depends very strongly on the meteorite in question. In the lowest petrographic types the matrix is a poorly understood fine-grained material that includes FeO-poor olivine, other silicates, metal and sulfide, spinel, chromite, and calcite. With increasing petrographic type it becomes coarser grained and fragmental, and eventually recrystallizes into grains of similar dimensions to the rest of the meteorite including chondrule grains (Christophe-Michel-Lévy, 1976; 1981; Ashworth, 1977; Huss *et al.*, 1981; Nagahara, 1984; Scott *et al.*, 1984; 1988; Taylor *et al.*, 1984; Nagahara and Kushiro, 1987; Brearley *et al.*, 1989; Hutchison, 1992; Brearley, 1996). The matrix also undergoes systematic composition changes – losing carbon and water and increasing in amount of metal as fine metal grains coalesce. Many authors consider the fine-grained matrix to be nebula dust, one author referring to it as the "holy smoke." It is thus the raw material from which chondrules were formed and Scott *et al.* (1984) actually show an image of a chondrule with matrix sealed inside. In other cases, authors have suggested that the matrix appears to have been produced from the chondrule mesostasis, that it is, in effect, crystallized fluid that was squeezed out of the chondrules. Of note, are that some chondrites contain unusual matrix, such as Bishunpur and Tieschitz, and that the different sorts of matrix could be formed in different ways.

Ordinary chondrite rims have been described by Allen *et al.* (1980), Rubin (1984a), Wilkening *et al.* (1984), Grossman and Wasson (1987), and Alexander *et al.* (1989). They are either fine-grained silicates or they are intimate mixtures of fine-grained silicate, sulfide, and metal grains. Some rims contain massive amounts of metal and sulfide. Most researchers think that they are essentially the same as the matrix.

4.5.3 Carbonaceous chondrites

The mineralogy and composition of the matrix in CI and CM chondrites has been described by McSween and Richardson (1977), McSween (1979a), Scott *et al.* (1988), Metzler *et al.* (1992), and Brearley (1993). Opaque minerals have been described by Kerridge (1964) and McMahon and Haggerty (1980). Rims have been described by Metzler and Bischoff (1996) and Bischoff (1998), who suggest that the matrix is, in effect, rims that run into each other. Metzler *et al.* (1992) also describe the rims, which they call "dust mantles" and argue that they formed in the nebula. Their evidence for aqueous reactions in the nebula is described in Fig. 3.5.

The matrix in CO chondrites has been described by Rubin and Wasson (1987a) and Keller and Buseck (1990). The matrix of Allende and its relationship to altered chondrules was decribed by Housley and Cirlin (1983), while the chondrule rims, referred to as "accretionary rims," were described by MacPherson *et al.* (1985).

4.5.4 Significance of rims

Hewins (1991) argued that the presence of rims suggests that the chondrules were free-floating in the nebula and that the rim is nebula dust accreted to the chondrules. This idea is consistent with the suggestion that chondrules formed in a dusty region of the nebula. Morfill *et al.* (1998) used the thickness of the rim and a model-dependent estimate of the rate of accretion to estimate the time chondrules were free-floating in the nebula. On the other hand, Huang *et al.* (1996a) suggested that the rims are recondensed chondrule volatiles that were generated during chondrule formation and condensed on the outside of the chondrule as it cooled. This idea was prompted by the observation that volatile-poor chondrules have thicker rims that volatile-rich chondrules. While we are still not certain about the origin of the rims and their relationship with fine-grained primordial matrix, these three works illustrate the potential importance of rims in understanding the accretionary history of chondrules and chondrites.

4.6 Stable isotope studies of chondrules

4.6.1 Oxygen

To a first approximation, chondrules from H, L, and LL chondrites have similar oxygen isotope compositions as their host meteorites, but they scatter so as to produce a single population (Gooding *et al.*, 1982; 1983). Chondrules from type 3 ordinary chondrites lie a few per mil to the right of the main chondrule field, along approximately a mass fractionation line, but there is still no indication that chondrules from the H, L, and LL chondrites are distinct from each other. Sears and Weeks (1983) and Clayton *et al.* (1991) have shown that loss of C present in type 3 ordinary chondrites as CO and CO_2 would move the chondrules from unequilibrated chondrites into the main field.

Chondrules from both ordinary chondrites and carbonaceous chondrites plot on lines with an approximate slope of unity but the carbonaceous chondrite trend turns toward the ordinary chondrite field at the heavy oxygen end (Fig. 4.13, Clayton *et al.*, 1983; 1991; McSween, 1985; Bridges *et al.*, 1998; Bridges, 1999; Li *et al.*, 2000). In some data bases there appears to be a dependence on chondrule size

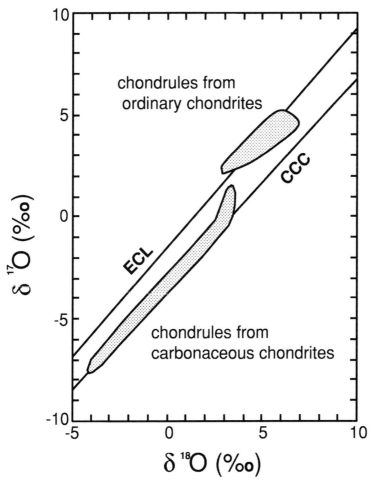

Figure 4.13 Three oxygen isotope diagram for chondrules. Chondrules from ordinary chondrites plot in a compact field on the "equilibrated ordinary chondrite line" (ECL) defined by ordinary chondrite whole rocks (termed "OC" on Fig. 3.4), while chondrules from carbonaceous chondrites plot on the line defined by anhydrous minerals from CV and CO chondrites (CCC) (termed "CCAM" on Fig. 3.4). While the slope of the two lines is approximately one, they intercept the graph to the bottom left at about $\delta^{17}O$ and $\delta^{18}O$ equal to -40 which might suggest a solid component with this highly ^{16}O-enriched composition mixing or exchanging with two gases whose composition plots at the top of the two lines. See caption to Fig. 3.4 for explanation of units (after Clayton et al., 1991).

(small chondrules having heavier oxygen suggesting exchange with oxygen near the terrestrial line) while in some cases mineralogy seems to be the major factor in the data because feldspar and cristabolite exchange more readily than olivine and pyroxene.

The ordinary chondrite and carbonaceous chondrite lines intercept at $\delta^{17}O$ and $\delta^{18}O$ equals $-40‰$. There has been considerable work relating the oxygen isotope

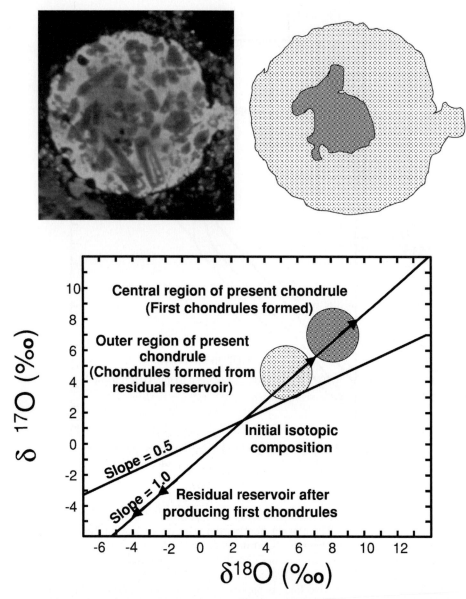

Figure 4.14 Top left, a cathodoluminescence (CL) image of a large group A1 chondrule from the Semarkona (type 3.0) ordinary chondrite. The olivine grains have red CL while the mesostasis is generally yellow but weaker in intensity in the central region. In the central region where the CL is low (top right), Na, K, and ^{16}O are depleted relative to the rest of the chondrule. Below, a figure from Thiemens (1996) adapted to show how these data can be explained by chondrule formation according to his laboratory experiments. The shaded circular regions correspond to data obtained in the similarly shaded regions in the map above. The central region of the original chondrule is ^{16}O and volatile element depleted, while the residual gaseous reservoir is higher in ^{16}O (along a slope-one line) and volatile element enriched. The outer region of the chondrule acquired ^{16}O (and volatile elements) by subsequent reaction between the chondrule and the residual gases (Sears et al., 1999a).

properties of chondrules to other properties, such as mineral composition, oxidation level, and size (Gooding et al., 1980; 1983; Clayton et al., 1981; 1983; McSween, 1985; Grossman et al., 1988; Huang et al., 1996b; Sears et al., 1997; 1998; Bridges et al., 1998; Bridges, 1999; Li et al., 2000).

Internal oxygen isotope variations have often been reported for type 3 ordinary chondrites (Mayeda et al., 1989), but only recently have they been observed for single objects in these meteorites. In one case, a large reduced chondrule in Semarkona (Group A) showing concentric zoning in many physical and chemical properties also showed zoning in oxygen isotopes (Fig. 4.14; Matsunami et al., 1993). Mesostasis at the center of the chondrule is isotopically heavier than oxygen at the perimeter and the data scatter along a slope-one line suggesting perhaps that processes responsible for the chemical zoning introduced the ^{16}O component. About 15% of the FeO-poor chondrules in Semarkona have compositionally zoned mesostases in which Na increases towards the outside (Grossman et al., 2002). These data suggest that the slope-one line observed in meteorites might have been produced by the event that created the chondrules.

The chondrules and refractory inclusions in the Allende CV chondrite have been especially well-studied because of interest in the inclusions as early condensates and because of their unusual properties. Chondrules and rims plot along with the bulk meteorite samples on the slope-one line, but with rims plotting apart from their host chondrules (Clayton et al., 1987). Most refractory inclusions also plot along the slope-one line, but several famous inclusions (known as C1, HAL, TE, CG14, and EK1–4–1) appear to have suffered a process causing mass fractionation (Fig. 4.15). These inclusions are quaintly called FUN inclusions, meaning "fractionated unknown nuclear."

The main question is how does the slope $\sim$1 line of the carbonaceous chondrite chondrules and inclusions relate to the slope $\sim$1 line of the ordinary chondrite chondrules? It has been suggested that the two lines are mixing lines between a solid of δ^{17}O and δ^{18}O equals -40‰ and two gases containing heavy oxygen isotope. However, Young and Russell (1998) recently showed that an altered zone on an Allende inclusion plotted on the carbonaceous slope $\sim$1 line while the interior plotted on the ordinary chondrite line. This suggests that the ordinary chondrite line is the more primordial and the carbonaceous line was produced by the alteration (Fig. 4.16).

4.6.2 Silicon, iron, potassium, boron, and hydrogen

Silicon isotopes have been measured in ordinary chondrite chondrules and Allende chondrules and lie on a mass fractionation (slope $\sim$0.5) line suggesting that there

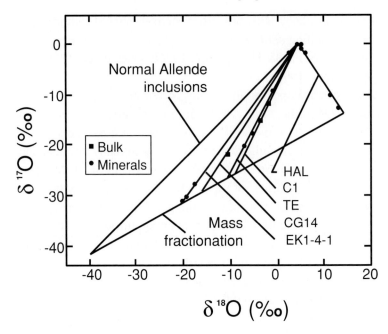

Figure 4.15 Oxygen isotope compositions for refractory inclusions in chondrites. Normal inclusions in the Allende CV3 chondrite plot on a line with a slope equal to one, but five unusual (so-called FUN (for "fractionated unknown nuclear")) inclusions plot on what appear to be mixing lines. At one end of the mixing lines is a point on the slope-one line at the top of the triangle in the figure, while the other ends of the lines plot on a mass fractionation line with a slope ~0.5 that intercepts the slope-one line at $\delta^{17}O$ and $\delta^{18}O$ equal to −40 (Lee et al., 1980).

was evaporative loss of silicon during the chondrule formation process (Clayton et al., 1983; 1985).

Potassium, iron, and boron isotopes have been measured in chondrules in an attempt to determine whether appreciable evaporation occurred during their formation or whether their properties are the result of mixing different precursors (Alexander et al., 2000; Alexander and Wang, 2001; Hoppe et al., 2001). If partial evaporation occurred, then the residue should be enriched in the heavy isotope, the degree of enrichment being calculable with the Rayleigh equation:

$$\frac{R}{R^o} = \left(\frac{X_1}{X_1^o}\right)^{(\alpha-1)} \tag{9}$$

where R is the present isotope ratio ($^{57}Fe/^{56}Fe$, $^{41}K/^{39}K$, $^{11}B/^{10}B$), R^o is the initial ratio, X is the concentration of the more abundant isotope now and X^o the original concentration, and α is the fractionation factor. In other words a 12% loss of an element should result in 3‰ increase in the heavy isotope in the residue while a 99.9% loss should result in an increase of 200‰. In fact, normal iron, potassium, and boron were found in the chondrules suggesting that either evaporation was

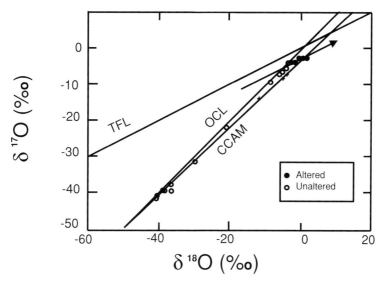

Figure 4.16 Three-isotope plot for oxygen showing the positions of the terrestrial fraction line (TFL), the line representing anhydrous minerals from the Allende CV carbonaceous chondrite (CCAM), and the line for ordinary chondrites (OCL). The data points refer to UV laser ablation in situ analyses of inclusion USNM 3576–1. Open circles represent unaltered parts of the inclusion; solid circles represent altered parts of the inclusion that occur along its edge. The arrow is the mass fractionation trajectory between the two. These data suggest that the primary Solar System line is that going through the ordinary chondrite data and that alteration produced the CCAM line (Young and Russell, 1998).

complete for these very volatile elements or, alternatively, normal potassium, iron, and boron reentered the chondrule after formation. Thus these particular tests for open-system behavior were inconclusive.

Hydrogen isotopes have been measured in chondrules by Hinton *et al.* (1983), Robert *et al.* (1987), and Morse *et al.* (1988). The values can be very high, in the range observed in the interstellar medium and ascribed to ion–molecule reactions, but it is unclear whether there is a correlation with class of chondrule or chondrule size. There is some evidence that the hydrogen lies on the surface of the chondrule or in the chondrule rim. Grossman *et al.* (2002) found that it correlated with Na in chondrules with compositionally zoned mesostases and had entered the chondrules during aqueous alteration.

4.7 Radiogenic isotope studies of chondrules

4.7.1 Xenon

There have been several studies of Xe isotopes in chondrules (Swindle *et al.*, 1983a; 1983b; 1986; 1991; Swindle and Grossman, 1987; Goswami *et al.*, 1998; Gilmour

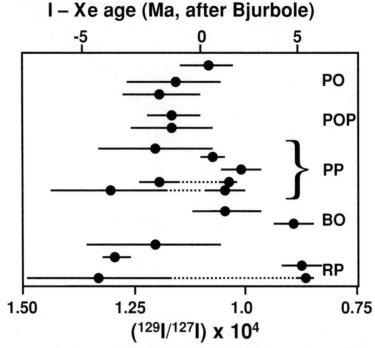

Figure 4.17 The ages of chondrules determined from the abundance of ^{129}Xe from the decay of ^{129}I, assuming that the initial distribution of ^{129}I was uniform (Swindle et al., 1983b). On this assumption, which seems increasingly likely, chondrules formed over a relatively short time interval of about 10 million years. The letters PO, POP, BO, PP, and RP refer to chondrule description: porphyritic olivine, porphyritic olivine–pyroxene, barred olivine, porphyritic pyroxene, and radiating pyroxene, respectively.

et al., 2000), all coming to essentially the same conclusion, namely, that chondrules became closed systems to ^{129}Xe from the decay of ^{129}I over a ~10 million year period several million years after the oldest Solar System solids formed (Fig. 4.17). Most authors accept this as the time of formation of the chondrules, although some argue for isotopic heterogeneity and others argue that the event actually being dated is the time when these chondrules reacted with water on the meteorite parent body. However, since other isotopic systems yield similar results it seems almost certain that it is the time of chondrule formation that is being dated.

4.7.2 Aluminum

A similar conclusion is suggested by ^{26}Al dating (Hutcheon et al., 1989; Russell et al., 1996; Misawa and Fujita, 2000; Huss et al., 2001; Mostefaoui et al., 2002). If it is assumed that the ^{26}Al distribution in the early Solar System was uniform, as

Table 4.6 *Distribution (number %) of chondrule groups in primitive meteorites in the various chondrite classes.*[a]

	A1	A2	A5	B1	Total
CI	0	0	0	0	0
CM	9	0	0	1	10
CO	30	few	few	5	35
CV	30	few	few	5	35
H	20	5	few	50	75
L	20	5	few	50	75
LL	20	5	few	50	75
EL	0	75	0	few	75
EH	0	75	0	few	75

[a] Estimates based on naked eye observations of cathodoluminescence mosaics by the author. The chondrules in EL and EH chondrites are essentially pure pyroxene while A2 chondrules in the other classes contain pyroxene although it is not always the major phase.

seems increasingly likely, then the chondrules formed over a several million year interval several million years after the oldest materials in the Solar System.

4.7.3 Chromium

Similarly, several authors have reported excess ^{53}Cr in chondrules and refractory inclusions due to the decay of now-extinct ^{53}Mn (half life 3.7 million years). In chondrules from ordinary chondrites the data suggest an initial ^{53}Mn/^{55}Mn ratio of ~9.5 × 10^{-6}. This corresponds to a time interval of 5.8±2.7 million years after the formation of the CAI (Birck and Allegre, 1985; Nyquist *et al.*, 2001).

4.8 Interclass comparisons

Dodd (1981) suggested that while chondrules in ordinary chondrites were primarily of the "lithic" sort, chondrules in carbonaceous chondrites were primarily of the "droplet" sort. Scott and Taylor (1983) took issue with Dodd, pointing out that there were similarities between carbonaceous, ordinary, and enstatite chondrites in terms of the Ca–Fe compositions of their olivines. To some extent the type of chondrules found in the various chondrite classes are very similar, but the proportion of chondrule groups varies from class to class. Table 4.6 is a first-order attempt to describe the chondrule populations of the different groups, although much detail remains to be determined, such as the similarity of the chondrules found in enstatite chondrites to the group A2 chondrules found in other classes. But in as much as the chondrules are pyroxene-rich they are superficially similar (Hutchison, 1982; Gooding, 1983).

4.9 Refractory inclusions

In the heady days before the return of Apollo samples from the Moon, a major meteorite shower occurred in Mexico. The meteorite, Allende, a CV chondrite, contained ~10 vol.% of unusual white inclusions consisting of highly refractory minerals, like gehlenite ($Ca_2Al_2SiO_7$) (Davis and MacPherson, 1988; 1996; MacPherson et al., 1988). It was not the only or even the first CV chondrite, but it was abundant and came at the right time when resources were being prepared for the return of the Apollo Moon rocks. Such inclusions, also known as CAI (calcium–aluminum-rich inclusions), have been found in trace amounts in most chondrite classes (Bischoff and Keil, 1983; Krot et al., 2001), however they are very abundant in CV chondrites and account for the high Ca/Si value for the class (Larimer and Wasson, 1988a).

When they were first observed it was suggested that these refractory inclusions, like chondrule formation (Marvin et al., 1970), were the residues of an evaporation process. However, in 1972 Grossman published a paper on the first stages of condensation and for many years the detailed properties of these inclusions were interpreted in terms of the condensation of the solids from the primordial solar nebula gases (Grossman, 1972). The idea emerged that the refractory inclusions were the first solids to form in the Solar System and dating generally seemed to confirm this hypothesis. When the stable isotopes in these inclusions were examined, they were also found to be unusual. As very curious material that may well be the first Solar System solids, the refractory inclusions have been intensely studied. But thermodynamics is about equilibria. It is not dependent on the process by which equilibrium is reached. Thus it remains possible that, rather than being the first condensates, these inclusions could be the residues of early evaporation events.

The refractory inclusions are subdivided on the basis of mineralogy and texture into types A, B, and C (MacPherson et al., 1988). Type A are fluffy, fine-grained inclusions that have suffered secondary alteration. Type B are larger, coarse-grained inclusions with little or no alteration. Type C are thought by most authors to be solidified melts, and thus effectively highly refractory chondrules. Other types of inclusion are identified by descriptive names such as amoeboid–olivine aggregates, spinel–pyroxene aggregates and spinel–hibonite–perovskite aggregates. All inclusion types are observed in CV chondrites, and all but type B are observed in CO and CM chondrites (although type A are rare in CM chondrites). Some of these refractory inclusions are found rarely in ordinary and enstatite chondrites.

The $^{26}Al/^{27}Al$ ratio of the refractory inclusions is about the highest measured (about 5×10^{-5}), making these inclusions the oldest Solar System solids. The related plagioclase-rich chondrules seem to be somewhat younger (Srinivasan et al., 2000) and the normal (sometimes called mafic) chondrules in unequilibrated ordinary chondrites seem younger still.

Everything about these inclusions reflects a refractory nature, including their metal grains (Fegley and Palme, 1985). Mo and W depletions can be reproduced by equilibrium calculations involving high-temperature gas–solid reactions, but require oxidizing conditions – a situation reminiscent of chondrules.

Geochemically, the most interesting properties of the refractory inclusions are their rare earth patterns. The rare earth elements, Yt to Lu, are refractory elements that are therefore abundant in these inclusions and have very similar chemical and physical properties. Yet they vary systematically in atomic radius and they show valence changes as electron shells fill, so they provide unique insights into the history of the inclusions. The refractory inclusions have been sorted into geochemical classes on the basis of rare earth element patterns (Mason and Taylor, 1982) and the geochemical history of these classes of inclusions has been determined (Boynton, 1975).

Like most chondrules, the refractory inclusions were formed in an environment far more oxidizing than any canonical solar nebula (Ihinger and Stolper, 1986; Kozul et al., 1988).

As with chondrules, refractory inclusions have rims that are thought to have particular genetic significance (Wark and Lovering, 1982; Bunch et al., 1985). Again as with chondrules, the nature of the rims depends on the object on which the rim appears, suggesting that the rims are the result of alteration processes. Typically the rims are 20–50 µm thick and consist of concentric zones of spinel–perovskite (innermost), melilite, or its alteration products, pyroxene and hedenbergite (outermost).

4.10 Relationship between chondrules and refractory inclusions

Intermediate between refractory inclusions and chondrules in many properties, most notably their bulk composition, are the olivine-bearing CAI, such as the "Fo-bearing CAI" in Allende (Wark et al., 1987), "poikylitic–olivine inclusions" in Allende (Sheng et al., 1991), "plagioclase-rich chondrules" in reduced CV chondrites (Krot et al., 2002), and the "anorthite-rich chondrules" in CH and CR chondrites (Krot and Keil, 2002). Thus the refractory inclusions are not compositionally discrete objects in meteorites, but are end-members of a compositional range with volatile-poor refractory inclusions at one end and volatile-rich chondrules at the opposite extreme. It seems possible that the whole range is an evaporation sequence, but some mixing of refractory and volatile materials may also have occurred (Misawa and Nakamura, 1996).

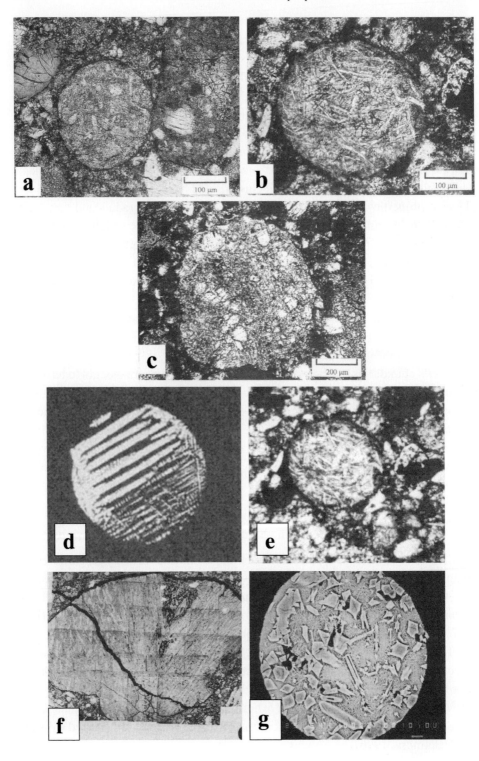

Many carbonaceous chondrites contain xenoliths frequently referred to as dark inclusions (Fruland *et al.*, 1978; Bunch and Chang, 1980; Johnson *et al.*, 1990). They appear to be a different type of carbonaceous chondrite closely related to the known classes and to the carbonaceous clasts in the ureilites.

4.11 "Chondrules" from other planetary bodies

A large number of authors have described crystallized lunar spherules (CLS), commonly referred to as "lunar chondrules" or "chondrule-like objects," in regolith breccias and soils from the Apollo collections and in lunar meteorites (Fredriksson *et al.*, 1970; 1973; Roedder, 1971; Bunch *et al.*, 1972; Dence and Plant, 1972; Keil *et al.*, 1972; King *et al.*, 1972a; 1972b; Kurat *et al.*, 1972; 1974; Nelen *et al.*, 1972; Roedder and Weiblen, 1977; Winzer *et al.*, 1977; Bischoff *et al.*, 1987; Warren *et al.*, 1990; Koeberl *et al.*, 1991; Neal *et al.*, 1991; Holder and Ryder, 1995; Symes *et al.*, 1998; Ruzicka *et al.*, 2000; Fig. 4.18a–f). They are widely considered to be impact-melt spherules, showing all the criteria designed to distinguish impact spherules from volcanic spherules (Delano, 1986; 1991). These objects have highland compositions, even when found in mare-rich soils, and are thus exotic to the site of the host rock. In their size distribution and abundance they resemble the chondrule population in CM chondrites. Their textures require long flight times (~20 minutes) consistent with an exotic origin. In fact, ballistic calculations suggest that an impact capable of producing a transient crater comparable in diameter to the radius of the target object is required by such flight times (Symes *et al.*, 1998; Ruzicka *et al.*, 2000). The Apollo rocks best known for their CLS are the Apollo 14 regolith breccias that are possibly Imbrium ejecta. The Imbrium basin has a diameter comparable to the radius of the Moon.

Figure 4.18 Chondrule-like objects (crystal-bearing spherules) in Apollo lunar samples and non-chondritic meteorites. (a) Segment of a radiating pyroxene chondrule in a silicate inclusion in the Netschaevo iron meteorite. (Longest dimension 1 mm, crossed Nicols). (b) A 250 µm type X spherule (with euhedral feldspar grains) in Apollo 14 highland regolith breccia 14315. (c) A 325 µm type Y spherule (with feldspar lathes) in 14318. (d) Spherule with barred texture from Apollo 11 soil. (e) Spherule similar to those in the Apollo 14 breccias from Apollo 11 soil. (f) Back-scattered electron image of a 130 µm spherule with skeletal crystals of olivine in the Kapoeta howardite meteorite. (g) Back-scattered electron image of a radiating pyroxene spherule (longest dimension is 4 mm) in the Pavlovka howardite meteorite. Image (a) is from Olsen and Jarosewich, 1971; (b) and (c) are from Symes *et al.*, 1998; (d) and (e) are from King *et al.*, 1972a; 1972b; (f) and (g) are from Olsen *et al.*, 1989.

Olsen *et al.* (1989) described six chondrule-like objects and brown glasses in howardites that they argued were soldified impact-melt spherules (Fig. 4.14f, g). They had textures indistinguishable from meteoritic chondrules, but Fe/Mn ratios that indicate they were indigenous and not simply contaminants from chondritic meteorites (Sears *et al.*, 1995c). Brownlee and Rajan (1973) have previously described similar spherules in howardites whose surfaces were microcratered.

Graup (1981) and Sears *et al.* (1996b) have described spherules from the suevite of the Ries Crater in Germany. These objects indicate that size-sorted glassy spherules can be produced by impact although in the case of the Ries impact they are rare. Ballistic calculations and laboratory experiments suggest their lack of crystallization is the result of very fast cooling rates for this relatively small impact onto a large object. In addition to glass spherules, these authors described accretionary objects somewhat analogous to the fine-grained rims on chondrules suggesting that the rims on chondrules might also be the result of processes occurring during impact.

5
Theories for the origin of chondrules

5.1 Some general comments

The data gathered on chondrules and chondrites over the last two hundred years are rich and diverse. There have been evolutions in techniques and instruments and large amounts of time and energy have been spent on the study of these data. Despite this, there is uncertainty as to the origin of chondrules and chondrites and this is reflected in the great many theories for their formation. This wealth of theories and lack of consensus on any particular theory is an indication of the complexity of the processes, the rudimentary state of our knowledge about conditions in the early Solar System, and, perhaps, the absence of some key element of information about the objects themselves.

Grossman (1988) reviewed chondrule formation theories and summarized them in table form, listing 19 theories. He concluded that chondrules formed in the nebula by an unknown flash heating event. Hewins (1989) presented a very different review to that of Grossman (1988), but came to essentially the same conclusion. Boss (1996b) presented a discussion of nine chondrule formation theories (that they are impact melts, meteor ablation products, the products of a hot inner nebula, FU Orionis, bipolar outflows, nebula lightning, magnetic flares, or that they were produced in accretion shocks, or nebular shocks). He gave an uncritical listing of the pros and cons of each. On the other hand, chondrule researcher J. N. Grossman once observed in a private communication that he had counted over 60 theories for chondrule formation. In this case the situation for the reviewer is analogous to that of an actor giving an acceptance speech after receiving an Oscar, afraid to start a list of thanks for fear of the inevitable omissions. Then there is the question of when a theory is sufficiently different from existing ideas to warrant fresh consideration. However, when examined in detail, theories can be sorted into two main groups with similar underlying ideas that can be discussed without going into the almost infinite number of second-order variations.

Thus we divide theories into those involving processes occurring in the dispersed dusty and gaseous primordial solar nebula, such as mega lightning or interstellar fireballs, and processes occurring on the meteorite parent body, namely the parent asteroid, such as volcanism or impact. Fortunately, the key issues then amount to whether the chondrules formed in the nebula by flash heating nebular dust, or whether they formed as impact melts on asteroid surfaces. There have been perceived problems with both scenarios, and this has resulted in variants of these ideas. In the case of nebula theories the main problems concern the mineralogical and chemical properties of the solids, so modifications to the "canonical" (i.e. astrophysically most reasonable) solar nebula are proposed. In the case of asteroid impact the main problem concerns energetics, so melted targets or even volcanism on the parent body are invoked. We will later argue that some of the perceived problems that have led to such a variety of ideas may not have stood the test of time.

5.2 Processes occurring in the primordial solar nebula

Proposed chondrule formation processes that might have occurred in the primordial solar nebula are shown schematically in Fig. 5.1. The ideas shown are not an exhaustive collection, but are meant to demonstrate the type and variety of ideas that have been discussed. Almost infinite variation on these ideas is possible. What they have in common is that they all involve a method for flash heating material in the nebula before it became part of an object larger than about a centimeter. We will discuss the ideas in more-or-less chronological order.

5.2.1 Solar activity

Sorby (1877) noted the absence of "Pele's Hair" (fibrous glass characteristic of volcanic glass) and suggested that chondrules were not volcanic but were the result of a molten spray being blown into an atmosphere at their fusion points, like "drops of fiery rain." This reminded him of solar prominences that were then a popular topic of research in the nascent science of astrophysics, so that the chondrules were described as "residual cosmic matter not collected into planets." The theory has similarities to a recent idea of Boss and Graham (1993). These authors observed that rapid changes in stellar luminosity, spectral lines, and light reflected from circumstellar nebulae are widely interpreted as due to large masses of material being accreted by the stars. They therefore suggested that chondrules were formed as this material fell into the star and was ejected as sprays of liquid droplets.

Liffman (1992) suggested that a 400 m object within 0.1 AU of the Sun exposed to solar gases streaming out from the Sun at <25 km/s produced chondrules by liquid

5.2 Processes in the primordial solar nebula

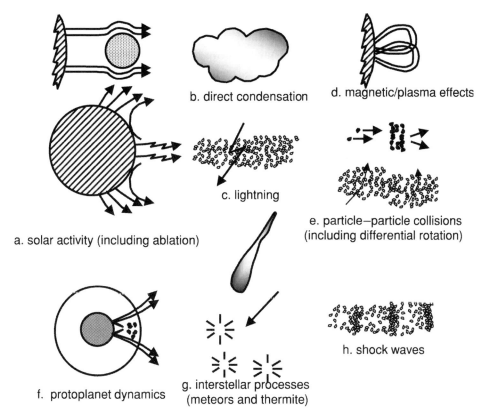

Figure 5.1 Some representative ideas of chondrule formation in the primordial solar nebula. (a) Several authors have suggested that chondrules are ejecta from the Sun, being formed near the Sun or from solar material (Sorby, 1877; Boss and Graham, 1993; Shu et al., 1996). Liffman (1992) suggested that chondrules are liquid run-off from a 400 m object within 0.1 AU of the Sun. (b) Wood (1963) suggested that chondrules were formed by direct condensation. (c) Whipple (1966) suggested that chondrules could have formed when lightning discharges passed through dust in the early solar nebula. (d) Sonett (1979) and Levy and Araki (1989) have proposed mechanisms involving magnetic fields and plasma. (e) Lange and Larimer (1973) suggested that chondrules were formed by particle–particle collisions in the nebula, which would be especially important near orbital resonances (Weidenschilling et al., 1998), and Rasmussen and Wasson (1982) suggested that chondrules were formed by friction due to differential rotation of zones of nebula dust. (f) Podolak and Cameron (1974) suggested that solids falling from the outer to the inner parts of the Jovian protoplanet produced chondrules as impact melts that would escape and move to the Asteroid Belt. (g) Wood (1983; 1984) suggested that chondrules formed from interstellar dust as it entered the primordial solar nebula, while Clayton (1980a; 1983) suggested that unstable radicals and reactive intermediates would undergo exothermal reactions as the grains warmed upon entering the inner Solar System. (h) Hood and Horányi (1993) proposed that shock waves in the primordial nebula produced chondrules.

Keplerian veolcity in the Asteroid Belt of about 20 km/s. Thus we require lightning flashes 10^5–10^6 times as energetic as on Earth and simple Keplerian motion cannot supply this. However it might be enough to create ionization that would enhance particle densities by a pinch effect, or, Whipple suggested, the lightning strokes could occur in the dusty atmosphere of an asteroid.

Cameron (1966) suggested that this mechanism explains the restricted size distribution of chondrules, since larger objects would not melt and smaller objects would completely evaporate. He calculated that it took three years for the chondrules to cool, presumably because the required gas and dust densities were so high.

It is worth noting that while most authors cite Whipple's suggestion that chondrules formed by lightning in the nebula, his concept of "the nebula" included the dusty "atmosphere" of an early asteroid. He added that such lightning could also stimulate the synthesis of organic prebiological compounds in the manner of the laboratory experiments of Miller and Urey discussed earlier. The idea that the chondrules may have been formed by lightning in the nebula has subsequently been pursued by many authors (Rasmussen and Wasson, 1982; Wdowiak, 1983; Horányi *et al.*, 1985; Eisenhour and Buseck, 1993; Eisenhour *et al.*, 1994; Gibbard and Levy, 1994; Love *et al.*, 1994; Wasson and Rasmussen, 1994; Horányi and Robertson, 1996).

5.2.4 Magnetic fields

Meteorites appear to have experienced very intense magnetic fields during their history, 0.1 to 10 G (Butler, 1972; Levy and Sonett, 1978; Sugiura *et al.*, 1979; Sugiura and Strangway, 1988). This has given rise to the suggestion that magnetic fields may have played a role in producing chondrules. Sonett (1979) suggested that solar flares moving along magnetic field lines in the midplane of the primordial solar nebula produced chondrules.

Levy and Araki (1989) suggested that instead of the magnetic flares originating in the Sun, there may have been wide magnetic field lines along which plasma moved into the corona of the nebula. Where lines intersected, there would be a high-stress region that would explosively release plasma, the energy release (E) being:

$$\frac{dE}{dt} = nL^2 \frac{B^2}{8\pi} \qquad (13)$$

where t is time, B is magnetic field strength, L is the width of the flare, and n is the rate at which the magnetic field collapses. The temperature (T) is given by:

$$T = \left(T_o^4 + \frac{B^3}{15\pi^{5/2} \sigma \sqrt{\rho}} \right)^{1/4} \qquad (14)$$

where T_o is the ambient tempearture, σ is Stefan's constant, and ρ the gas density. To melt a 10 mg chondrule thus required a magnetic field of 10 G and very low gas density (10^{-18} g/cm). Levy and Araki (1989) and Fleck (1990) debated the feasibility of these treatments and discussed Sonett's (1979) suggestion that chondrules could have been formed by magnetic solar flares in the nebula midplane. However Levy and Araki (1989) thought that there was too much neutral gas for the flares to penetrate.

5.2.5 Impacts between grains

Lange and Larimer (1973) first suggested that chondrules might be formed by impacts between dust grains in the primordial solar nebula. A couple of years later Kieffer (1975) suggested that droplet chondrules were formed by jetting during impact in a regolith. However, these ideas were questioned by Kerridge and Kieffer (1977) who argued that since lunar chondrules coexist with abundant agglutinates that are absent in primitive meteorites, then chondrules could not have been formed by impact into a regolith. Instead, they suggested that the jetting previously described by Kieffer (1975) occurred by particle-to-particle collisions in the nebula.

A problem with impact between dust grains in the primordial nebula is that the energy of the impacts will generally be too low to melt the grains. Weidenschilling *et al.* (1998) circumvented this by arguing that orbital resonances with Jupiter might have caused dust near the resonances to increase the dispersion in their orbital elements and undergo particularly violent impacts that resulted in the formation of chondrules. Similarly, shock waves passing through the dust in the primordial solar nebula could have caused compression and violent interactions between dust grains that resulted in fusion and chondrule formation. Attempts are being made to take this theory to ever greater degrees of quantitative detail (Hood and Horányi, 1991; 1993; Ruzmaikina and Ip, 1995; 1996; Desch and Connolly, 2002). A variant of this idea is that grains interacted by turbulent shear beneath the nebula surface (Wood, 1986).

5.2.6 Impact within the Jovian protoplanet

Podolak and Cameron (1974) suggested that solids falling from the outer to the inner parts of the Jovian protoplanet would produce impact melts that would escape and move to the Asteroid Belt. By having impact inside the protoplanet the pressure was high enough to produce liquids and O/H ratio could be 25–30 times cosmic because much of the hydrogen had been removed to produce the Jovian core.

5.2.7 Interstellar processes

Wood (1983) suggested that clumps of interstellar dust could have undergone heating as they entered the primordial solar nebula in much the same way as meteors entering the Earth's atmosphere – first they were heated as they passed through a shock front and then as they underwent drag heating. The process was treated quantitatively by solving equations for the conservation of mass, energy and momentum, and aerodynamic drag. The final temperature of the dust aggregates was given by:

$$T_p = \left(\frac{\alpha \rho_g v^3}{B\beta\sigma}\right)^{1/4} \tag{15}$$

where T_p is peak temperature, α an accommodation coefficient (0.5), ρ_g is the gas density, v is the velocity of the dust aggregate, B is the surface area of the particle, β is the thermal emissivity (0.75) and σ is Stefan's constant. For ρ_g equal to 3.1×10^{-11} g/cm^3 and a velocity of 56.6 km/s, a 1 mm aggregate reaches 1450 K in 16 s. Differences in infall velocity, gas density, amount of silicates and ice, and time traversing a proposed dust-rich region just inside the shock front would determine whether the object formed would be a CAI or a chondrule, and the size and reduction state of the chondrules.

An alternative heating mechanism for interstellar grains entering the solar nebula was proposed by Clayton (1980b; 1983; 1988), based on earlier work by Urey and Donn (1956) and Greenberg (1976). These authors suggested that interstellar grains forming at 50 K would contain considerable energy in the form of radicals and reactive intermediates that would be highly unstable, and upon heating would undergo exothermal reactions that would release enough energy to melt the grain. Amorphous mixtures of MgO, SiO, and FeO, for instance, would release 340 kcal/mol as they crystallized into enstatite and iron and would entirely evaporate if not for radiative cooling. The grains might also trap various amounts of oxygen, to give a variety of redox states in the final chondrules, and aggregates rich in Ca and Al (that Clayton calls SUNOCONS) could form CAI.

5.2.8 Aerodynamic drag heating

If rapidly moving dust grains in the primordial nebula suddenly encountered a pocket of gas, they might become heated by aerodynamic drag. Certainly, there is evidence for something similar to aerodynamic sorting in the chondrule populations (see Section 1.6). The relevant relationship can be obtained by equating energy loss by radiation with energy gain by aerodynamic drag heating:

$$\varepsilon 4\pi r^2 \sigma r T^4 = \pi r^4 \, {}^1\!/_2 \, \rho_{gas} v^3 \tag{16}$$

where ε is the object's emissivity, r its radius, T is its temperature, and v is its velocity; σ is the Stefan–Boltzmann constant, and ρ_{gas} is the density of the gas. Several authors have proposed this idea or developments from the idea (Wood, 1984; Hood and Horányi, 1991; Hood and Kring, 1996; Scott et al., 1996) and were aware of the difficulties in obtaining the required pressures and densities for the nebula gas.

5.2.9 Shock waves

Hood and Horányi (1991; 1993) proposed that shock waves in the primordial nebula produced chondrules. Essentially, these authors computed the velocity needed for dust particles to move through a gas in order to bring the particles to the melting point, so this proposal is essentially another means of aerodynamic heating. Thus the temperature can be derived from the appropriate energy balance equation while velocity will be determined by the drag forces on such particles:

$$m_{\text{d}} C \frac{dT_{\text{d}}}{dt} = 4\pi a^2 q_{\text{d}} + 4\pi a^2 \varepsilon_{\text{abs}} J_{\text{r}} - 4\pi a^2 \varepsilon_{\text{em}} J_{\text{r}} \sigma T_{\text{d}}^4 \qquad (17)$$

$$m_{\text{d}} \frac{dV_{\text{d}}}{dt} = \pi a^2 \frac{C_{\text{D}}}{2} \rho_{\text{g}} V_{\text{d}}^2 \qquad (18)$$

where m_{d}, T_{d}, C, a, and V_{d} are the dust grain mass, temperature, specific heat, radius, and relative velocity, respectively, q_{d} is the gas–dust heat transfer rate per unit surface area, J_{r} is the external radiation energy flux, and ε_{abs} and ε_{em} and are the dust absorption and emission emissivities. Finally, C_{D} and ρ_{g} are the drag coefficient and the density of the gas. Hood and Horányi find that Mach numbers of >3.1 bring a small fraction of the dust to the melting point while Mach numbers of >4 melt most of the grains.

5.3 Processes occurring on parent bodies

5.3.1 General comments

Proposed chondrule formation processes involving impact or volcanism that might have occurred on meteorite parent bodies are shown schematically in Fig. 5.2. Again this is meant to be illustrative of the range of ideas discussed, rather than being exhaustive, but it is probably the case that the variants proposed are far fewer than in the case of mechanisms occurring in the nebula. It should be mentioned that many of the theories discussed in the previous section could apply to the dusty and perhaps gaseous environment of a newly accreting asteroid or a volatile-rich asteroid (say, ~ 20 vol.% water like CI chondrites) during a global-scale impact.

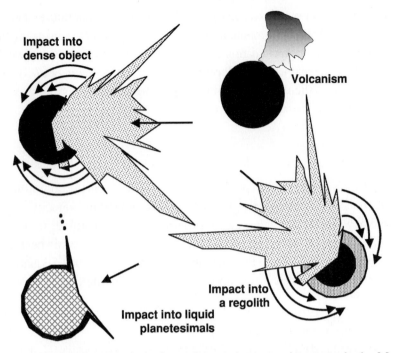

Figure 5.2 Ideas for chondrule formation on the meteorite parent body. Many workers in the 1950s and 1960s (e.g. Urey and Craig, 1953) assumed that chondrules were impact-melt spherules, but when it was thought that impacts were not energetic enough to produce melts it was suggested that the target was a molten asteroid (e.g. Zook, 1980; Sanders, 1996). More recently, it has been suggested that impact into a loosely consolidated highly porous object would produce considerable impact melt that would be retained on the surface (Sears *et al.*, 1996b).

5.3.2 Volcanism and volcanism-related processes for chondrule formation

It is of interest to consider some of the nineteenth and early twentieth century thoughts on meteorite genesis and how so much was established so early. Partly because this helps prevent us from reinventing the wheel, but also because it helps detach us from currently popular but probably transient ideas. More importantly, in the face of so much modern data and ideas it is sometimes helpful to go back to first impressions to help focus on the main features to be explained.

Merrill (1920) summarized much of the early work. Reichenbach (1860) suggested that each particle (chondrule) was an independent meteorite made by breaking up a preexisting stone. Sorby (1864) was probably the first to note that "globules" had "crystallized after solidification," while Tschermak (1885) also remarked on the "tuff like" character of meteorites and wrote: "The variability of composition of chondrules; the occurrence in them of glass inclusions, the occurrence of glass chondrules; indented chondrules; and joined chondrules – all argue that chondrules are not solidification structures peculiar to magnesium silicates, which formed in the body of the chondrite, but that the chondrules are rapidly cooled liquid drops, many of which, being extremely brittle, were broken." Haidinger (1867) considered chondrules as abraded rock fragments of the sort often found in volcanic eruptions and explosions. Daubree (1879) also thought that attrition was important in chondrule formation and the idea persists today (Dodd, 1981; Kimura and Watanabe, 1986). Later, Borgstrom (1904) observed that the diversity of chondrules is such that they must have formed by solidification as separate units before being accumulated into a common rock. He also noted undulose extinction and metallic veins that post-dated accumulation and indicated that "the stone was an integral part of a larger body," and "meteorites are fragments of dispersed planets." Brezina (1885) suggested crystallization from a magma. Klein (1906) likened chondrules to spherulites and Wahl (1910) thought that chondrules formed when silicate melts cooled in a heated atmosphere. Fermor (1938) suggested that chondrules were remelted garnets from the upper mantle of a parent body.

Merrill (1920) was perhaps the first to appreciate the significance of the distinction between porphyritic and non-porphyritic chondrules, suggesting that the former were abraded melt fragments and the latter were fused drops of fiery rain, that "of the many varietal forms of chondrules not all may have similar origin." He also likened chondritic texture to that of some furnace slag where metal spherules are encased in glass.

Mason (1960) suggested that chondrules and metal would segregate in a gravity field so he proposed that a CI-like chondrite underwent dehydration followed by reduction in the presence of carbon:

$$Mg_4Fe_2Si_4O_{10}(OH)_8 \xrightarrow{600°C} 2(Mg, Fe)_2\underset{\text{olivine}}{SiO_4} + 2(Mg, Fe)\underset{\text{pyroxene}}{SiO_3} + 4H_2O$$

$$\xrightarrow{C} 4\underset{\text{pyroxene}}{MgSiO_3} + 2Fe + 2CO_2 \tag{19}$$

Mason suggested that this occurred during close solar passage, and he noted that future developments in geothermometry might enable this to be tested. The paper was criticized by Urey (1961).

Similar views to those of Mason were later developed by Ringwood. Ringwood (1959) accepted the arguments of Lovering *et al.* (1957) that the body on which chondrites formed was large. Noting that although chondrites were "tuffaceous and fragmental in origin," Ringwood rejected the then-current consensus that chondrules were "drops of molten silicates which have been chilled." Instead, he argued that on large bodies water-saturated magma would rise to the surface and form chondrules in one of three ways: by spherulitic crystallization (previously suggested by

Klein, 1906); by rapid precipitation of Mg-rich silicates following loss of Fe at the surface; and by "cavitation" (rapidly crystallizing viscous melts trapped in steam bubbles). After supporting these ideas with test tube experiments, Fredriksson and Ringwood (1963) suggested that "the eruption (of water-bearing volatiles to the surface) may form a very mobile gas–solid system as an ignimbrite or *nuée ardante*." Fredriksson (1963) also favored explosive volcanism on a C-rich body, arguing that impact would not produce multiminerallic chondrules, and that direct condensation would not produce the high-FeO content of olivines in primitive chondrites.

5.3.3 Impact onto a parent body

The second half of the twentieth century began with Urey and Craig's (1953) paper pointing out the chemical distinction between H and L chondrites. In discussing chondrule origins, Urey and Craig (1953) pointed out that while the escape velocity of the Moon is much higher than that of a 100 km asteriod (2.3 km/s, compared with 140 m/s) the Keplerian veolcity in the Asteroid Belt (19 km/s) is sufficiently high that a 10% dispersion in velocities, due to variations in eccentricity and inclination, would result in impacts sufficient for melt production. They also pointed out that a 10 km object impacting a 100 km object at 3.7 km/s would result in the fragmentation of the larger object, explaining the rubble-like structure of many meteorites. In such a scenario, the chondrules were assumed to be condensed droplets from the impacts, or impact-melt spherules.

Urey (1967) favored a lunar-sized meteorite parent body, arguing that liquid droplets destined to become chondrules could not be gravitationally retained by asteroidal sized objects. He remarked that, "the lack of enthusiasm for a collisional origin of chondrules is primarily related to the enthusiasm for an asteroidal origin of meteorites." Nevertheless, efficiency was a problem in making chondrules by impact into a lunar-sized body, and why did the melted ejecta form spherules rather than a large melt pool? Maybe there was impact into the melt pool. So he suggested that after multiple impacts a low thermal conductivity surface would increase the amount of heat deposited and thus chondrule formation, and that as the impact rate declined brecciation and shock would result rather than chondrule formation.

Petrographic properties of chondrules suggest to some petrologists that they were once part of a "planetary" object (Hutchison *et al.*, 1988; Hutchison, 1996). Wlotzka (1969) likened the shock-melted metallic spherules in the Ramsdorf chondrite to chondrules, and since the metallic spherules were known to have been heavily shocked to the point of partial melting suggested that meteoritic chondrules were poorly shock-melted droplets. Dodd (1971; 1974) and others (Dodd and Teleky, 1967; Dodd and Van Schmus, 1971) also thought that shock melting played an important part in chondrule formation. He suggested that the different chondrule types showed signs of being shocked to differing levels and proposed that chondrules were formed by shock processes during accumulation of dust onto asteroid-sized bodies. His "metal-poor microporphyritic chondrules" were fragments of impact melts that had been abraded during subsequent low-velocity impact, his "metal-rich microporphyritic chondrules" were shock-melted during impact with some loss of volatiles, and the "excentroradial and barred" chondrules were quenched droplets that had suffered only light to moderate shock.

Kieffer (1975) pointed out that subsequent sorting of chondrules would not have been necessary if droplet chondrules were produced as jets as impact melts broke up. If the surface is porous then particle-to-particle collisions with jetting into pore space could have occurred in the regolith. Clastic chondrules, which are more numerous than droplet chondrules, were still assumed to be erroded fragments.

Urey and Craig (1953) mentioned the idea that chondrules could have been formed by impact into impact-melt sheets, but with the discovery that ^{26}Al was active during the early phases of Solar System history, the possibility exists that chondrules formed by impact into entirely molten parent bodies. Zook (1980) suggested that a <20 km molten object with a sintered surface was impacted, "perturbations by the still growing proto-Jupiter may be important in putting many objects on collision trajectories," and would produce a spray of small objects that would reaccrete after decay of the ^{26}Al. The process would have high efficiency, produce little fragmentary material, and not produce agglutinates. Zook (1981) added that the energy required is about 6000 erg/g (equivalent to an impact velocity of about 1 m/s). Off-center impact between two hot planetesimals was suggested by Smith (1982) and Leitch and Smith (1982) as a way of explaining the coexistence of enstatite grains with red and blue cathodoluminescence (CL) in the same chondrule. They pointed out that enstatite with blue CL was approximately consistent with fractional condensation while enstatite with red CL was approximately consistent with near-equilibrium condensation under highly reducing conditions. Chondrule formation by collision of molten objects was pursued in greater depth by Sanders (1996) who also presented thermal models and addressed several detailed criticisms by Taylor *et al.* (1983) and Grossman (1988).

6

Discussion of theories for the origin of chondrules

6.1 The primordial solar nebula and possible cosmochemistry

6.1.1 Physical conditions in the early nebula

Reviews of the primordial solar nebula from which the Sun and planets formed have been published by many authors (Safronov, 1972; Herbig, 1978; Cassen and Boss, 1988; Weidenschilling, 1988; Wood and Morfill, 1988; Boss, 1996a; Cassen, 1996; Hartmann, 1996). Essentially, a rotating mass of gas and dust with the same composition as the Sun and similar stars underwent gravitational collapse to form a rotating disk of gas with dust (Morfill et al., 1993; Dubmile et al., 1995; Fig. 6.1). The gravitational collapse was at free-fall velocities for much of the process so that the formation of protoplanets or planetary embryos occurred very rapidly, within 10^5 years according to some estimates. Accretion of interstellar material occurred throughout this process, so as time went on the central core competed with the disk to accrete new material.

The central core evolved into the proto-Sun which underwent various energetic processes as it established equilibrium between gravitational collapse and the internal pressure due to nuclear reactions in the core. Bipolar outflows and periods of stellar instability were associated with this early phase of Solar System history (Hartmann and Kenyon, 1985; Clarke et al., 1990; Sauer, 1993; Liffman and Brown, 1996).

Fragmentation of the disk produced subclouds within which dust accreted into planetesimals (e.g. Cameron and Fegley, 1982; Cuzzi et al., 1993; 1996) and the planetesimals accreted into planets. However, resonant gravitational interactions with Jupiter and Saturn prevented the accretion of planets in the region of the Asteroid Belt. In fact considerable mass was ejected from the Asteroid Belt during this process, perhaps in the form of Mars-sized protoplanets (Chambers and Wetherill, 1998).

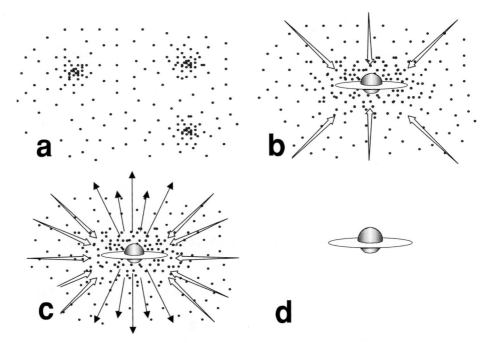

Figure 6.1 The four stages of star formation. (a) Grains and dust accumulate into stellar cores as the cloud fragments and the energy to sustain the cloud is lost. (b) The individual core collapses to form a protostar surrounded by a circumstellar disk. Collapse occurs from the inside out. (c) A stellar wind breaks out along the rotational axis of the protostar, causing a bipolar flow. (d) At the end of infall, a star is formed, surrounded by a circumstellar disk from which planets will form. (Shu *et al.*, 1987; 1993).

The great challenge of Solar System formation is explaining the angular momentum distribution – that while most of the mass of the System is in the Sun, most of the angular momentum is in the planets. In other words, some means of transferring momentum from the Sun to the planets is required and while there are a great many ideas, some very promising, the cause of the angular momentum distribution remains the big uncertainty in understanding Solar System evolution. This is the astronomical setting against which chondrules and chondrites formed.

In order to understand chondrules and chondrites, the crucial properties are the composition, pressure, and temperature of the nebula, as well as the plethora of physical processes likely to have been occurring. Composition seems fairly straightforward. It seems fairly clear that the composition of the nebula was the same as that of the Sun's photosphere which is also the same – at least for the condensible elements – as the primitive chondrites. The oxygen fugacity (chemical activity, essentially the same as pressure) is given by equation (6) and shown in Fig. 3.1

(Section 3.1.2). Conditions are highly reducing because of the preponderance of hydrogen.

Astronomers have always assumed that the primordial solar nebula was a very cold place, with temperatures similar to those observed in molecular clouds. Only within a few tenths of an AU of the Sun would temperatures be significantly above 400 K. On the other hand, the meteorite data have been used to argue for a hot nebula since meteorites contain evidence for high-temperature events, not just chondrules but also the formation of refractory grains and other high-temperature components that produce the different Mg/Si and Ca/Si ratios of the various chondrite classes (Morfill, 1983; Boss, 1988; 1993). However, these discussions assume that these meteorite properties were the result of equilibrium nebula chemistry rather than being stochastic non-equilibrium nebula effects or parent body effects.

Finally, it is important to know the pressure of the nebula. Most estimates suggest that the pressure of the nebula in the vicinity of the Asteroid Belt was 10^{-5} to 10^{-7} atmospheres. The densities of the planets and equilibrium nebula chemistry have also been used to determine pressure–temperature curves assuming adiabatic conditions (Lewis, 1976a; 1976b; Fig. 6.2). This suggests temperatures of ~700 K and pressures of $\sim10^{-3}$ bar in the vicinity of the Earth, about ~320 K and 10^{-4} bar in the Asteroid Belt, and ~200 K and 3×10^{-5} bar at the orbit of Jupiter.

6.1.2 Chemical processes in the early nebula

The work of Lewis (1976a; 1976b) has been very influential in our thinking about Solar System chemistry, and equilibrium chemistry has been helpful in thinking about the composition of the outer planets and that of their atmospheres. The presence of volatility trends in the chondritic meteorites, astrophysical ideas on the formation of the Solar System, and compositional trends in the planets, led to the idea that the chemical history of the solar nebula could be calculated using equilibrium thermodynamics. It was assumed that the nebula was entirely gaseous, that it was hot, and that its composition was equal to that of CI chondrites for most elements and the photosphere for the non-condensible elements. With these assumptions it is then possible to predict the sequence of elements that would have condensed out of the gaseous nebula (Lord, 1965; Larimer, 1967; Grossman, 1972; Grossman and Larimer, 1974; Anders, 1977; Wai and Wasson, 1977; Sears, 1978).

The results are only partially successful in explaining chondrite properties. For example, the chondrules in the ordinary and carbonaceous chondrites are more oxidized than can be explained in this way and the enstatite chondrites contain minerals like TiN, CaS and Si-bearing metal, that require the conditions to be a factor of two more reducing than cosmic abundances allow (Kelly and Larimer, 1977). Nevertheless, this process of condensation does provide a framework against

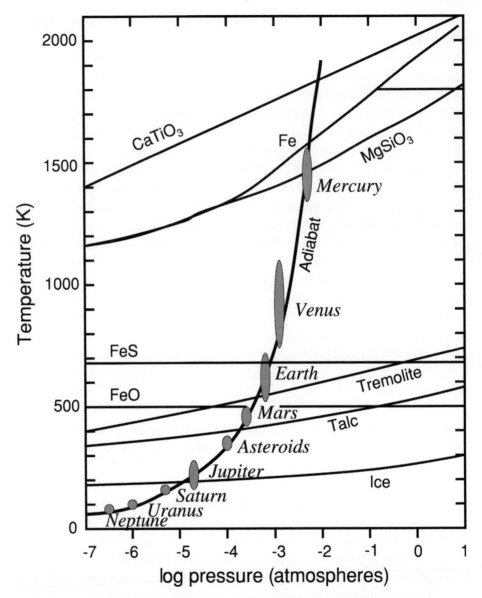

Figure 6.2 Pressure–temperature curves at which the major chemical species are stable compared with a reasonable adiabat for the early Solar System (Lewis, 1976b).

which to explain many chondrite properties and by-and-large they are reasonably successful. For example, it is possible to explain their volatility trends (Cassen, 1994; 1996; 2001), why Mg/Si patterns vary from one class to another, and why the rare earth elements have certain abundance patterns in refractory inclusions. The first phases to condense are – as expected – rich in refractory elements and can be likened to the refractory inclusions. The next wave of condensation produces solids with a higher average ratio of Mg to Si than later condensation, so maybe it is the addition or removal of these components that produces the different Mg/Si ratios of the different classes.

However, explanation is not proof, and there is always a nagging doubt that other explanations might be possible. Equilibrium thermodynamics calculates the phases that are at equilibrium with each other – it does not explain how those phases came to exist. Perhaps the condensation sequence is actually an evaporation sequence and the refractory inclusions, instead of being the first condensates, are actually very old evaporation residues, as Marvin *et al.* (1970) originally suggested. We could still imagine that it was the removal or addition of high-Mg/Si components that created the different Mg/Si ratios in the classes, but those components were also evaporation residues rather than condensates. The fact that astrophysicists believe that the nebula was never hot might lead us to favor this second idea, but we still need to identify the mechanism that caused the heating.

This could be where the chondrules come in. The one thing we know about chondrules is that they were once very hot. They were heated to the point of being molten. Bournon (Howard, 1802) referred to them as "curious globules" inferring that perhaps he thought they were once molten, and when Sorby first saw these objects through the microscope he described them as rapidly cooled droplets. In other words, chondrules have been known to be solidified melt droplets from the day they were first seen. Some of them were entirely molten, some were only partially molten, but all have been close to the melting point. It could well be that it is the chondrule-forming process that made the refractory inclusions, produced the Mg/Si differences between classes, and gave rise to the volatile element trends, at least for the moderately volatile elements whose abundance does not vary with petrographic class. In their case, their abundance patterns are somehow related to parent body metamorphism. The question is, how did chondrules form? What was the heat source?

6.2 Critique of nebula theories for chondrule formation

6.2.1 Some background

There is a curious geographical factor influencing belief in whether chondrules formed in the nebula or whether they formed by a process acting on the parent asteroids. In North America there is an inclination towards theories that had chondrules form in the nebula, although there is little or no agreement on which theory. Throughout the rest of the world there is no preference for nebular theories and there may be a preference towards parent body mechanisms such as meteorite impact.

The reasons for this geographical difference are a matter for the historian, but such trends in opinion are not uncommon.

Researchers favoring a nebula origin for chondrules will cite a series of null arguments against their formation by impact. One review, for example, spent five and a half pages rejecting impact theories, four pages rejecting other non-nebula theories, and only three-quarters of a page describing nebular theories. The arguments against an impact origin go something like the following.

- The impact-melt spherules on the Moon are rare and compound spherules are unknown. This suggests that impact into a regolith is not sufficiently efficient to produce chondrules.
- Impact into the lunar regolith produces far more agglutinates than impact-melt spherules, and yet agglutinates are absent in chondrules.
- Impacts comminute and yet comminuted material is absent from chondrites.
- Hypervelocity impact pits are common on lunar grains but are either absent on chondrules or show a completely different morphology.
- Chondrules have the same ancient ages as the chondrites whereas impact should have been occurring throughout the age of the Solar System.
- Impact-melt spherules on the Moon or on Earth are embedded in original target rock, yet there is no prechondrule target rock in chondrites.
- The production of chondrules by impact into a regolith should result in chondrules "rolling around in a dusty regolith" and they should therefore have uniform composition. However, rims on meteorite chondrules vary in composition from chondrule to chondrule.
- Impact melts tend to be uniform in composition yet meteoritic chondrules vary in composition.
- Suevitic melts contain impact-melt spherules but they also contain angular melt clasts.
- Asteroids are small, with escape velocities only a few hundred meters per second. Thus any impacting object would not impart sufficient energy to generate impact melts that could become the soldified droplets we call chondrules.
- Even if impacts were energetic enough to make melts, the melted material would be the most energetic material in the target and would entirely overcome the weak gravitational field of the asteroid and escape into space.
- Kozul *et al.* (1988) have also argued that impact produces more rapid cooling rates than observed for chondrules.

However, null arguments are dangerous, because they are difficult-to-impossible to prove. Proofs require understanding, and lack of a proof can be a result of poor understanding. And certainly we have had little or no understanding about the nature of the asteroids and the mechanics of impact onto asteroids. The Moon and the Earth have large gravity fields and everything that happens near their surfaces is heavily influenced by this; asteroids are essentially microgravity environments and the physics there is very different. The first insights into the nature of asteroids, at least their surfaces, are only just beginning to appear as a result of spacecraft exploration of asteroids and this will surely lead to a revision of these theories and arguments.

Nebular theories are numerous (see Chapter 5), and it is difficult and not particularly helpful to discuss each one in turn because they are often very speculative, poorly constrained, and they draw on such a breadth of technical ideas. The truth is that their very diversity and lack of agreement on any one of them suggests that we have not found the right answer. Furthermore there is compelling evidence from the chondrules themselves against the nebular theories. This is described below. However, nebular theories share enough in common that a critique of them collectively is adequate. The following discussion is based on papers by Sears *et al.* (1995a), Symes *et al.* (1998), and Ruzicka *et al.* (2000).

6.2.2 Chondrule density in the Solar System

Chondrules are very abundant in the most famous and largest class of meteorites, namely the ordinary chondrites, where about 75% by volume of these meteorites is composed of chondrules. If this was typical of the chondrule population in the early solar nebula, then the amount of energy involved in the chondrule forming event was enormous, in fact comparable to the gravitational energy of the Solar System (Levy, 1988). It would be as if nature had found a way of funneling all the energy available to it into chondrule formation. We would be looking for a process that was Solar System in scale and universal.

However, it is almost certainly not the case that the ordinary chondrite population is typical of the chondrule population of the Solar System as a whole. Ordinary chondrites are termed "ordinary" because they are abundant on Earth, not because they are abundant in space, and the ordinary chondrites – and therefore the chondrules – are grossly overrepresented in the terrestrial meteorite collection. Chondrites are tough and it is only the tough meteorites that can survive passage through the Earth's atmosphere. Calculations show that a meteorite with the mechanical properties of a CI or CM chondrite undergo 1000 times as much fragmentation as a meteorite with the density of an ordinary chondrite (Fig. 6.3). Many objects – more fragile than the CI and CM chondrites – are probably entering the atmosphere and not surviving the strains of atmospheric passage. For over forty years the US military has been detecting these objects. They give rise to the famous "airbursts." We actually know of several instances of objects entering the atmosphere producing very large spectacular fireballs and depositing very little, if any, meteoritic material on the Earth. Furthermore, when material is deposited it invariably turns out to be CI chondrites or other similarly fragile meteorites.

There is evidence that although there are a great many ordinary chondrites, they actually come from very few asteroids and very few breakup events. The cosmic ray

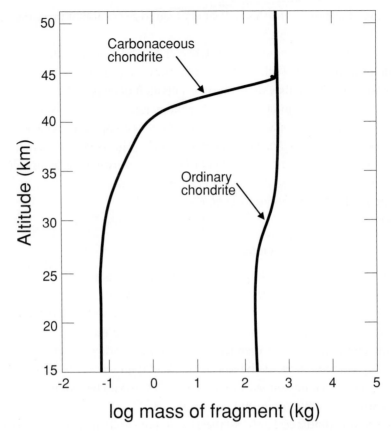

Figure 6.3 Graph showing the mass of a meteorite surviving atmospheric passage as the meteorite passes through the atmosphere. The curve for two meteorites are shown – a very tough ordinary chondrite that undergoes little fragmentation and a very soft carbonaceous chondrite that shatters and disintegrates as it passes through the atmosphere. In general, the more primitive the material the more fragile it is, so the terrestrial meteorite flux is biased towards the stronger less primitive materials (Baldwin and Shaeffer, 1971).

exposure ages of the H chondrites and LL chondrites (Fig. 2.5) and Ar–Ar ages of the L chondrites (Fig. 2.6) indicate that a significant fraction, perhaps the majority, of meteorites from these classes were produced by the same fragmentation event. Thus a few relatively recent catastrophic breakups sent large numbers of meteorite fragments to Earth from two or three parent bodies.

Finally, the spectra of sunlight reflected from the asteroids does not match that of the abundant ordinary chondrites, but resembles the spectra of classes that are either rare on Earth or unknown (Fig. 2.3). As discussed earlier (Chapter 3), it might be that "space weathering" is disguising the true nature of the asteroid surface, but it could also be that ordinary chondrite material simply is rare in the Asteroid Belt. Space weathering must occur to some extent. It is impossible to imagine that an asteroid surface would not be altered in some fashion by millions of years of exposure to space, but the diversity of surfaces still visible might suggest that the initial differences are surviving space weathering and providing an indication of the true nature of the subsurface materials.

Thus the process that formed the chondrules might sometimes be highly efficient, but it might happen only rarely (Sears, 1998). In fact, the type of chondrite that is usually thought to most closely resemble the original starting material, and now resembles the Sun's photosphere, is the CI chondrites that do not contain any chondrules. The class that comes second after the CI chondrites in their "primitiveness" is the CM chondrites, and these contain only $\sim$10 vol. % chondrules. Thus the classes that probably best represent primitive Solar System material are actually quite poor in chondrules. The mechanism for producing them need not be Solar System wide. It might be highly localized and very efficient.

6.2.3 Chondrule sizes

In addition to knowing the number densities of chondrules and their sometimes high levels of oxidation, we know that the chondrules have been size sorted, i.e. each of the chondrite classes has its own characteristic range of chondrule sizes. This is also true of the size of the metal grains and this will be discussed further below. In the nebula, gas densities are such that particle size sorting by aerodynamic drag would require distance scales comparable to the dimensions of the Solar System (Weidenschilling, 1977). This is possible, but at the same time other properties, like the complementary compositions of chondrite components and the elemental and isotopic interactions, require much higher gas densities than would be found in the solar nebula. Scott and Haack (1993) have discussed the chemical fractionations of chondrites in terms of aerodynamic sorting.

6.2.4 The existence of compound chondrules

Compound chondrules have been known since the earliest chondrule studies and demonstrate that sometimes chondrule densities were high enough that one chondrule could impinge on another while still plastic. Sometimes one chondrule is indented in the side of another, and sometimes one chondrule is completely enclosed in another (Wasson, 1993). Usually the two chondrules are of similar texture, suggesting similar composition and history, in other words similar maximum temperatures, nucleation site density, and cooling rates. It is clear that one or both of the compound chondrules were fluid droplets when they collided.

Most chondrules had cooled completely when they impacted the surface or were otherwise incorporated into the debris that was to become the accreted chondrite. However, a few instances have been reported of chondrules that were still plastic when they were agglomerated into the rock so they have crystal grains and fragments pressing into them. It is clear from these observations that at the place of formation, chondrule densities were very high and that the time spent in free flight was not always long enough to ensure solidification before accumulation. The sites of chondrule formation and chondrite formation were close in time and space. This is more consistent with the theory of chondrule formation by impact on the parent body; also it is not necessarily clear that the formation of chondrules in the nebula would result in these properties.

6.2.5 Oxygen fugacity of chondrules

Another lesson to be learned from chondrules is that they did not go through their melting process in contact with a gas that in any way resembled our current concepts of the solar nebula. This is a conclusion that no researcher has challenged. The oxygen content of the ambient gas must have been enriched over the cosmic value by factors of about 1000. It seems very likely that oxygen was not the only element to be so enriched, but other elements like sodium appear to have entered the chondrule while it was still hot.

Several researchers have proposed that an increase of the dust-to-gas ratio by a factor of ~1000 would produce the required oxygen content because oxygen is present in the dust as silicates but hydrogen is virtually all concentrated in the nebular gases. Of course, the dust would include other lithophile elements like sodium. Other methods of increasing the reactivity of oxygen can also be imagined. For example, the use of supersaturation, which increases the reactivity of gaseous iron with oxygen by delaying the condensation of iron (Blander and Katz, 1967). Thus all of the mechanisms for producing chondrules in the solar nebula that have been proposed also require major departures from nebular conditions in order to increase the amount of oxygen and other lithophile elements.

6.2.6 The "complementary" composition of components

We have suggested that chondrules formed by an efficient localized mechanism because only a few related classes of chondrites contain significant numbers of chondrules and that these are not the most primitive meteorites. But there is evidence that the chondrule forming process is even more localized than this. One of the most significant points about chondrite compositions is that although the chondrules, metal, sulfide, refractory grains, and matrix differ from each other in composition, when mixed together and the whole rock analyzed their composition is remarkably close to that of the Sun (Wood, 1985). It is as if the elements had come together in solar proportions and major processing had occurred to produce the different components, but that they were never sufficiently separated to lose the starting composition of the original material. From an analytical point of view this is very fortunate because relatively small sample masses are needed to obtain a representative sample of a meteorite. Some analysts argue that 10 g is all that is required, while others have argued that as little as 0.25 g is all that is required to capture all of the components in their correct proportions. This is a convenience for analysts and curators, but a major scientific conclusion here is that this homogeneity of composition argues for a highly localized process. Chondrules and their associated metal and matrix were not separated from each other by more than a fraction of a centimeter or if they were they were very well mixed with a thoroughness that extended to the subcentimeter scale and this also argues against formation in the nebula.

6.2.7 Chondrule cooling rates

It is not just the size and composition of the chondrules that argues against an origin for the chondrules in anything remotely resembling the canonical solar nebula. It is their physical history too – chondrules cooled very slowly. In the words of one author, it is as if the chondrules were bathed in a warm bath as they cooled, rather than being exposed to free space. Chondrule cooling rates are on the order of 1–1000 °C/h, while a chondrule-sized object in free space would cool at $\sim 10^6$ °C/h (Grossman, 1988). Apparently, chondrules were located close to other warm material while they cooled. Chambers and Cassen (2002) calculated that in order to obtain observed chondrule cooling rates, 10–100 km sized collections of nebular dust are required. A 10–100 km clump of dust moving through the early Solar System might be regarded as an asteroid in the making, just as envisaged by Whipple (1972a; 1972b)!

6.2.8 Charged particle tracks

We have shown that chondrule composition is not as we would expect given formation in the solar nebula, and there were major departures from equilibrium. So what else can we do to seek evidence that chondrules existed as independent entities in free space?

Allen et al. (1980) searched for charged particle tracks on the surfaces of chondrules. Solar particle radiation is of such a low energy that it can only penetrate a few micrometers into the surface grains on an astronomical object (Fig. 6.4). Similarly, micrometeorites impact grains on the surface of airless bodies and produce microcraters on the surface. Both solar radiation damage tracks and microcraters are observed on grains from the very surface of the Moon. A chondrule floating

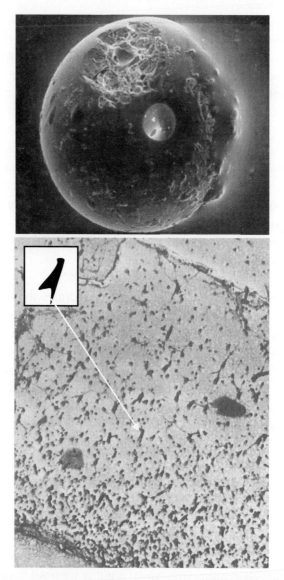

Figure 6.4 Microcraters and charged particle tracks on the surface of rocks exposed directly to space. Upper figure, a 250 μm glass spherule from the Apollo 11 soil containing a microcrater in the upper left corner of the figure (NASA photograph). Lower figure, scanning electron microscope image of a plastic replica of an acid-etched olivine grain from lunar soil 12037 (Bhandari *et al.*, 1972). The dark objects (sketch inset) are the charged particle tracks and their density increases towards the edge of the grain. These tracks are typically 1 μm in length. Microcraters and charged particle tracks should be present on any silicate material exposed directly to space, but neither were observed on chondrules by Allen *et al.* (1980).

in free space should pick up radiation damage tracks and microcraters, yet Allen *et al.* (1980) found neither suggesting that the chondrules were never independent entities in free space. Others have rebutted this conclusion by saying that the nebula was too dusty for charged particles to pass through the nebula, but this is unclear, and it would not explain the lack of micrometeorite impact pits.

6.3 Formation of chondrules by impact into a regolith

All the evidence seems to suggest that chondrules did not form in the nebula as we know it. It seems sensible to return to the arguments that made us look at nebular theories in the first place, and we are reminded that these were largely null arguments; arguments based on the fact that we could not explain chondrule formation by impact onto an asteroid. But times have moved on. Since these null arguments became popular, robotic spacecraft have flown by and even orbited asteroids and we now have new data and ideas about their surfaces. We also have important new data on the chondrules, such as their formation ages. It is time to reconsider the old arguments against these enigmatic objects being the result of meteorite impacts onto an asteroid.

6.3.1 Impact velocities were not too low to make melts

It was originally assumed that chondrites formed when the Asteroid Belt was a peaceful place, with gas and dust moving around the Sun and coming together relatively gently. A few grains stuck to each other, grew in total cross section to capture other grain aggregates, and then eventually began accreting by gravitational forces. The gravitational forces on such small objects were weak, so that impact velocities were small, much too small for impact melting. Petrographic observations indicated that chondrule formation, and accretion, metamorphism, and brecciation of the chondrites overlapped in time and that that time was very short (Scott *et al.*, 1985).

But the picture has dramatically changed. Radiometric observations indicate that chondrule formation occurred up to ~4 million years after the formation of the first solids. Such data also indicate that some asteroids had become large (several hundred kilometer) bodies, some even with an active igneous geology (Hewins and Newsom, 1988). They were extruding basalts onto their surfaces. There were large asteroids in the Solar System before the chondrules formed. If chondrule formation was delayed by this much time, then the Asteroid Belt was not a quiet peaceful place, but it was violent with the mighty proto-Jupiter and proto-Saturn stirring up the Belt and increasing relative velocities of asteroids to about 6 km/s, more than enough to make impact melts (Podosek and Cassen, 1994; Russell *et al.*, 1996). Since Jupiter formed very quickly, within ~10^5 years of the onset of accretion (Cameron, 1995), the Asteroid Belt was "stirred up" by resonances with proto-Jupiter or the jovian core (Davis *et al.*, 1979), so that mean relative velocities were ~5 km/s prior to the formation of the chondrules and thus sufficient to produce impact melts. The early Asteroid Belt was a violent place with major impacts making craters, and even fragmenting the asteroids, and associated with every crater was impact melt. What happens to this melt on the asteroid is a matter for current research, but under conditions of microgravity and the violence of impact it presumably fragments into a million tiny droplets that would cool, solidify, and crystallize as they floated about the asteroid under microgravity conditions. The question is whether these droplets would leave or return to the asteroid.

6.3.2 Meteorite parent bodies were able to retain most of the impact ejector

Two decades ago it was usually assumed that a meteorite impacting an asteroid would produce fragments of ejecta that would have sufficient velocity to easily escape the asteroid (Housen *et al.*, 1979). But recent spacecraft images have led to a rethinking of this conclusion. Early observations of Gaspra indicated that it has a regolith (Belton *et al.*, 1992). Observations of Ida and Mathilde suggested the same, and now we have over a year's worth of observations of Eros that indicate a widespread and rather thick regolith. Apparently, small asteroids are capable of retaining deep regoliths. It would seem that asteroids are not solid rock that will transfer the impact energy to the ejecta and give it enough velocity to leave the asteroid. Instead, asteroids must be made of porous material that can absorb the shock, so the ejecta have little energy and can be retained. Thick regoliths can build up. Calculations suggest that for a typical impact on Gaspra, 50–70% of the ejecta (depending mainly on impact velocity) would return to the parent asteroid (Asphaug and Nolan, 1992; Housen, 1992).

Calculations indicate that so much of the ejecta is retained that included among it is a significant amount of melt. Melt droplets produced when the impact-melt sheet is splashed into space may float above the asteroid for a considerable period of time before descending to the surface (Fig. 6.5; Scheeres *et al.*, 2002).

6.3.3 Chondrule (and other component) sorting could have occurred on the meteorite parent body

Chondrule and metal grain size sorting could have occurred on the asteroid during the accretion phase when these asteroids were forming from nebula clouds of material. However, this seems unlikely because accretion occurred very quickly

6.3 Regolith impact origin of chondrules

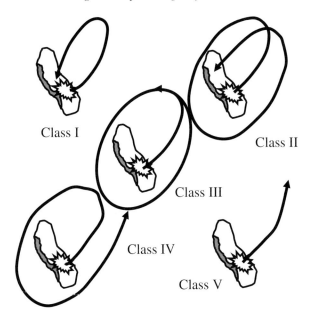

Figure 6.5 The five outcomes for ejecta of an impact on an asteroid according to Scheeres *et al.* (2002). Class I, immediate impact. Class II, eventual impact. Class III, long-term stable orbit about the asteroid. Class IV, eventual escape. Class V, immediate escape. Classes I and V are expected to be the most common outcomes; Class III is thought to be very unlikely.

(within about 100 thousand years) and chondrules are ~4 million years old. It is most probable that sorting occurred at the time of chondrule formation, when major impacts disrupted large fractions of the asteroid and not only made melt droplets but also produced large clouds of gas and dust.

We do not understand the internal structure of asteroids. We know that densities are low (except for the igneous asteroids with basalt surfaces) (Chapter 2). This is even true of asteroids with S classification that are assumed to have dry surfaces. Maybe they are very porous inside, with a considerable amount of void space. However, many asteroids have densities that are too low for porosity alone to explain and they must also contain considerable amounts of water, either as water-ice or as water of hydration. In fact, as discussed in Chapter 2, spectroscopic evidence for water has been found on a large number of asteroids.

The impact into such a volatile-rich target under microgravity probably resembles an exploding volcano more closely than it resembles a terrestrial or lunar impact. The evaporation of water and other volatiles from the regolith of the meteorite parent body would produce a temporary atmosphere that would "fluidize" the dust and create conditions suitable for aerodynamic and gravity sorting of chondrules. Large clouds of dust suspended and transported by gas are one of the most dangerous volcanic features and are called "*nuée ardentes*," meaning "glowing avalanche" (Fisher *et al.*, 1980; Valentine and Fisher, 1993). The flux and velocities of the fluids required to produce these processes on asteroids are both surprisingly small because of the size of the parent body. If chondrules formed in such a regolith, then the temporary atmosphere had an oxygen isotope composition near the terrestrial line on the three oxygen-isotope plot (Clayton *et al.*, 1991), similar to that of CI chondrites.

6.3.4 There are lunar chondrules on the Moon's surface

The lack of chondrules on the lunar surface is often cited as an argument against an impact origin for chondrules, yet lunar chondrules are present in Apollo 14 breccias in about the same abundance as CM chondrites ($\sim$10 vol. %). Simple ballistic calculations show that crystallized impact spherules require long flight times and that only craters comparable in size to the target produce ejecta with these long flight times. Thus most impacts on the Moon produce agglutinates or glass spherules (Symes *et al.*, 1998). As images of Gaspra, Ida, and Mathilde demonstrate, impacts producing craters with diameters comparable to the radius of the target can be important for certain asteroids and these will be the bodies where chondrules dominate the local regolith.

7

Discussion of theories for metal–silicate fractionation

We suggested in Section 3.3 that the two features that need to be explained in understanding chondrites are the formation of chondrules and the metal–silicate fractionation. Having understood, at least partially, these two processes, much of the remaining properties of chondrites will fall into place. We discussed chondrule formation in the previous chapter, now we discuss metal–silicate fractionation and how the chondrules and metal were assembled together to produce chondrites. This, we argue, amounts to discussing the origin of chondrites. Finally, we can examine the extent to which other chondrite properties relate to the ideas we have developed for these processes.

7.1 Chondrule sorting

We will consider several processes for sorting chondrules. First, we will consider the idea that size sorting is a primary property and a result of the formation mechanism. Second, we will discuss the idea that it is the passage of the chondrule through the local gas environment that sorts the chondrules. We will call this aerodynamic sorting. Next, we will discuss the idea that the sorting is a result of the process by which chondrules were made to move from their formation location to the location at which the present meteorite formed. We call this ballistic sorting. Finally, we suggest that abrasion could result in size sorting in some cases.

7.1.1 Primary processes

Most theories for sorting chondrules and metal grains are "secondary," in the sense that they assume that a range of sizes previously existed and were later sorted. There is an implicit assumption that chondrules in all classes came from a single size distribution curve. In fact, it is possible that the mechanism that originally produced the chondrules formed a different size distribution at each location.

7.1.2 Aerodynamic processes

Whipple (1972a; 1972b) suggested that a planetesimal moving through a cloud of dust and gas from which it was accreting chondrules would preferentially sweep up larger chondrules because of their greater resistance to movement through a viscous fluid as given by the Stokes force (F):

$$F = 6\pi \eta s v_s \quad (20)$$

where s is the particle radius, v_s the relative velocity of the planetesimal and cloud of gas and dust, and η is the viscosity of the gas ($\frac{1}{2}vL\rho$, where v is the mean kinetic speed of the gas molecules, L is the mean free path of the gas molecules and ρ is gas density ($>10^{-8}$g/cm^3 for the Stokes regime, or $\rho/3\rho_c$ for the Epstein regime where ρ_c is the gas density at which the gas molecule mean free path equals the radius of the asteroid)). Whipple then showed that 50 μm chondrules in Stokes drag would be accreted onto an asteroid of radius S km if the velocity v_1 was less than:

$$v_1 = \frac{0.36\eta}{s^2 \rho_s} S \quad (21)$$

Assuming η is 1.6×10^4 dyne s/cm^2 and ρ_s is 3 g/cm^3, then limiting velocities for accretion onto 0.1, 1, 10, and 100 km asteroids are 3.4, 34, 340 m/s, and 3.4 km/s, respectively, although equation (21) may breakdown for the largest body. The equivalent expression for Epstein drag is:

$$v_1 = \frac{0.11 v\rho}{s\rho_s} S \quad (22)$$

Clayton (1980b) proposed that aerodynamic sorting by grain size during condensation of gases in the interstellar medium was responsible for elemental fractionation patterns in stellar and supernova condensate CAI. Weidenschilling (1977) and Weidenschilling and Ruzmaikina (1994) have discussed broader aspects of the aerodynamics of solids in nebula and the coagulation of grains, but Wieneke and Clayton (1983) argue that the turbulent low-mass model nebula assumed by these workers could not produce grains larger than one centimeter diameter because drag causes the particles to spiral into the Sun.

7.1.3 Ballistic processes

Ballistic calculations are notoriously difficult and involve determining not just deceleration in the manner described in the previous section, but also the acceleration that launched the projectiles. In addition, there are a great many perturbations on both acceleration and deceleration. A detailed analysis of a possible role for ballistic processes on chondrule and metal grain sorting has not been made, although some calculations relevant to crystallized lunar spherule formation by lunar impact were described by Symes *et al.* (1998).

The ballistic trajectory for particles ejected from a source depends on their velocities, the gravitational field, their angle of flight, the presence and density of the atmosphere, and the mass of the particles. Therefore, all else being equal, a group of particles moving together will have their individual trajectories determined by their individual masses and thus size sorting will occur. Fragments ejected from a meteorite impact, particles ejected from a comet nucleus, or volcanic bombs ejected from a volcano, generally show the effect of sorting during acceleration and smaller particles travel further from their source than larger fragments. This is in contrast to aerodynamic sorting during deceleration, such as a shower of meteorites moving through the atmosphere, where larger fragments travel furthest along the line of travel and the "scatter ellipse" – well-known to meteorite recoverers – is the result. Cannonballs fired from cannons might be expected to show either behavior.

7.1.4 Abrasive processes

Several authors have suggested that certain chondrules, "lithic" chondrules, are abraded fragments of igneous rocks (Dodd, 1981; Kimura and Watanabe, 1986). This suggests another mechanism for size sorting for these chondrules, namely the duration and intensity of tumbling in an abrasive environment.

7.2 The metal–silicate fraction in the nebula

Urey and Craig (1953) and Anders (1964) assumed that the only way to separate metal from silicate on the parent body was to produce melts and gravitationally separate the immiscible fluids. They argued that the textures of chondritic meteorites clearly precluded this. Thus most theories for metal–silicate separation involve processes occurring in the primordial solar nebula (Fig. 7.1). Newsom (1995) continues to favor a nebula mechanism for metal–silicate fractionation.

7.2.1 Mechanisms involving differences in crystal growth

Donn and Sears (1963) pointed out that solids forming in the primordial nebula would grow by screw dislocations which produced long "whiskers" 1 μm or so in diameter. Because the whiskers would be strong yet ductile, they would be very efficient at trapping similar particles moving at high velocity. These objects would quickly accumulate into "lint balls" with high surface area to volume ratio (about 100 times that of a compact crystal) and very low densities (~ 0.1 g/cm^3).

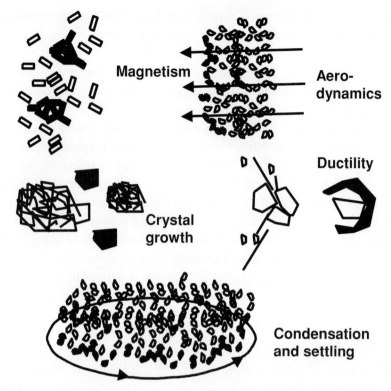

Figure 7.1 Some ideas for separating silicates and metal in the primordial solar nebula. Donn and Sears (1963) suggested that metal would accrete less rapidly because it would grow in the nebula as compact grains with small collision cross sections, while silicates would grow as "fluff under the bed" and thus accrete more readily than metal. Orowan (1969) suggested that the greater ductility of metal would ensure its preferential accretion. Dodd (1976) suggested that aerodynamic sorting would separate metal and silicate because of their differing sizes, while Larimer and Anders (1967) suggested that magnetism would ensure preferential accretion of metal grains. Finally, Larimer and Wasson (1988b) suggested that since metal and silicates have different condensation temperatures they would form solids at different temperatures and fall to the medium plane of the Solar System at different times.

The rate of crystal growth (dN/dt, the nuclei created per cubic centimeter per second) is related to the amount of supersaturation (α, the ratio of pressure to vapor pressure for the condensing molecule) according to:

$$dN/dt = B\exp(-A/T^3 \ln^2 \alpha) \qquad (23)$$

where A and B are constants and T is temperature (K); α should be much greater for metal than silicates, so metal grains would show a much greater tendency to form lint balls than silicates and by their differential coagulation metal and silicates would become separated.

7.2.2 Ductility-based mechanisms

Orowan (1969) suggested that the low density of the Moon could be due to the scavaging of iron by the Earth. He suggested that since iron is ductile at most temperatures, while silicates are brittle, iron would preferentially accrete by both hot and "cold" welding whereas silicates would need to be hot to weld together. Larimer and Anders (1967) applied this idea to chondrites and suggested that differential accretion by silicates and metal due to their different ductility could also fractionate silicates from metal in the Asteroid Belt region of the nebula.

7.2.3 Mechanisms involving magnetism

Larimer and Anders (1970) used siderophile element abundances to argue that magnetism was involved in separating metal and silicates. The atomic abundances of Ir, Pd, Au, Fe, Ni, Co, and Ge in L chondrites, as a ratio to their abundance in H chondrites, are 0.6–0.7, while for Ga and S these ratios are close to unity. According to Larimer's (1967) calculations, this meant that metal started to separate from silicates between 1050 K (the condensation temperature of Ge) and 985 K (the condensation temperature of Ga). This is similar to the Curie point of iron, 900–940 K. Thus, Larimer and Anders (1970) argued that magnetism was involved in separating metal and silicates. Harris and Tozer (1967) have argued that magnetism could enhance the capture cross section of metal grains in the primordial solar nebula by a factor of 20 000.

Two developments have weakened the Larimer and Anders argument for a magnetic effect in metal–silicate fractionation. First, the condensation temperatures for Ga and Ge have undergone considerable revision since 1967 (Sears, 1978a; 1978b), and no longer does the Curie point seem to play any special part in the process. Second, Ga and S are present in unmetamorphosed chondrites in non-magnetic phases; at low temperatures Ga appears to be lithophile and S is chalcophile. Since they do not condense in the metal, they do not provide information on the temperature of metal–silicate fractionation.

7.2.4 Mechanisms involving aerodynamics

Dodd (1967; 1976) took Whipple's treatment one step further by equating v_1 for coexisting metal and silicate. Then

$$s_m^x \rho_m = s_s^x \rho_s \qquad (24)$$

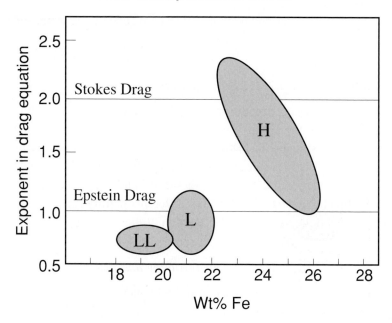

Figure 7.2 Plot of the exponent in the expression $\rho_m d_m^x = \rho_c d_a^x$ for coexisting metal and chondrules in ordinary chondrites (Dodd, 1976). An exponent of one corresponds to Epstein drag while an exponent of two corresponds to Stokes drag. The data for the ordinary chondrite classes not only fail to cluster around one of the theoretical exponent values, but also most of the data for L and LL chondrite are less than 1.0.

where the subscripts s and m refer to silicate and metal, and x is 1 for Epstein drag and 2 for Stokes drag. Dodd suggested that the drag regime changed with meteorite class and distance from the Sun and, following Whipple's arguments, proposed that the size of the chondrules in a given meteorite class reflected the size of the parent body (LL > L > H). In a similar fashion, Skinner and Leenhouts (1993) showed that chondrules and metal grains in the CR chondrite Acfer 059 had similar terminal velocities and thus proposed aerodynamic sorting in the nebula, arguing that the process happened radially rather than vertically because of the rarity of metal-rich meteorites. While the data are in the right ball park, they do not fit the theoretical predictions very well. Instead of clustering around an exponential coefficient of 1 or 2, depending on drag regime, they have values ranging from very much less than 1 to greater than 2, suggesting that another process is complicating the picture (Fig. 7.2).

7.2.5 Condensation and settling

Larimer and Wasson (1988b) have discussed an idea in which elements were initially in the gas phase and condensed down into solids as the temperature of the nebula

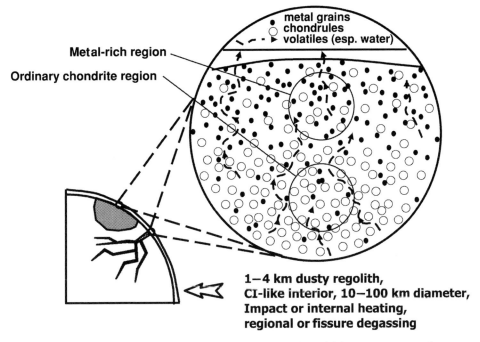

Figure 7.3 A scenario in which silicates and metal could become separated on the meteorite parent body (Huang et al., 1996a; Sears and Akridge, 1998). The body was assumed to be originally water-rich, much like CI chondrites or cometary nuclei; heating of the interior of this body would then result in the drying out of the surface layers and the passage of large volumes of water from the interior through these layers. Thus surface materials would be size and density sorted, detailed experiments indicating that the finer grained metal would be lifted through the chondrules to produce a metal-rich surface layer and a subsurface layer with chondrite proportions of metal and chondrules. The degree of metal–silicate separation depends on the size ratio of chondrules and metal grains.

cooled, spatially and temporally. Once formed, the solids would settle from their formation location in the nebula to the midplane. Thus there would be a spatial or temporal variation in the composition of the material settling to the midplane. Since metal is more volatile than silicates under some conditions, it is a relatively simple matter to imagine conditions that would explain metal–silicate fractionation. It is, of course, possible to explain any volatility-related elemental patterns by such mechanisms (Wasson and Chou, 1974; Wai and Wasson, 1977; Wasson, 1985).

7.3 Metal–silicate fractionation on the parent body

Huang et al. (1996a) have suggested that metal and silicate grains were separated by aerodynamic and gravitational sorting in a thick dusty layer, somewhat akin to a regolith, on the parent body as gas flowed from the interior of the volatile-rich

Table 7.1 *Calculated and observed metal abundances in H, L, and LL ordinary chondrite meteorites (Huang et al., 1996b).*

Flow rate v(mm/s)[a]	$x_{LL} : x_L : x_H$ (weight ratio)
$1.3 \times v_{mf}$	1.0 : 2.1 : 3.9
$1.2 \times v_{mf}$	1.0 : 2.7 : 6.8
$1.1 \times v_{mf}$	1.0 : 4.5 : 7.4
(observed)	1.0 : 3.0 : 8.0

[a] v_{mf} refers to the minimum flow rate of LL chondrules, calculated from the Ergun equation.

(CI-like) body (Fig. 7.3). Lange and Ahrens (1982) have discussed the formation of impact-generated atmospheres and Lebofsky *et al.* (1989) have reviewed the volatile inventory of asteroids. The situation can be described theoretically by the Ergun equation (Kunii and Levenspiel, 1991), which equates the upward drag force (which depends primarily on the Reynold's number, R_e) to the downward gravitational force:

$$\frac{1.75 R_e^2}{\varepsilon^3 \phi} + \frac{150(1-\varepsilon) R_e}{\varepsilon^3 \phi^2} = \frac{d^3 \rho_g (\rho_s - \rho_g) g}{\mu^2} \quad (25)$$

where ε is the void fraction under minimum flow conditions, ϕ the sphericity and d the diameter of the particles, ρ_g and ρ_s the densities of the gas (calculated assuming an ideal H_2O gas) and solids, μ the viscosity of the gas, and g is the acceleration due to gravity. R_e depends on the flow rate of the gases and is given by:

$$R_e = \frac{d v_{mf} \rho_g}{\mu} \quad (26)$$

where v_{mf} is the minimum flow rate required for fluidization, which is proportional to the acceleration due to gravity. The calculated velocities for fluidization on these small, asteroid-sized, objects are very low, -1, -10 and -100 mm/s for 10, 100, and 1000 km radius parent objects, respectively, and could be sustained for months to years assuming CI abundances of water in the interior of the object.

Segregation is dependent on particle density, particle size, and gas flow rate (Rowe *et al.*, 1972), maximum segregation being obtained when the flow rate is just above the minimum required for fluidization. The empirical relationships for the separation of particles of size d_b and d_s (where the subscripts refer to "big" and "small") and density of ρ_h and ρ_l (where the subscripts refer to "heavy" and "light"

7.3 Metal–silicate fractionation on the parent body

yield the combined expression (Rowe et al., 1972):

$$x = k(v - v_{mf(f)}) (\rho_h/\rho_l)^{-2.5} (d_b/d_s)^{0.2} \tag{27}$$

where $v_{mf(f)}$ is the minimum flow velocity for the flotsam (the material which floats, in this case the chondrules), which is given by the Ergun equation above. This relationship accounts well for the metal–silicate fractionation shown by the ordinary chondrites (Table 7.1).

8

So how far have we come and where do we go next?

8.1 Chondrules and chondrite classes as impact pyroclastics

If chondrules formed by impact into a regolith, and chondrules behaved as open systems during their formation, then the diversity of chondrule compositions presumably reflects the diversity in the intensity of impact. It is then a small step to assume that the redox state of the resultant chondrite similarly depends on the violence of impacts locally. The remaining factor in forming chondrites concerns the matter of assembling the components, and producing small variations in the amount of matrix and metal in relation to the chondrules. The size and distribution of the chondrules and metal, which are characteristic of many classes of chondrites, suggests sorting before or during accumulation. Again, a great many mechanisms have been proposed for how this might have been achieved in the nebula, but I think it unlikely that this process occurred in the nebula because the meteorites managed to preserve compositions so close to cosmic and because aerodynamic sorting alone fails quantitatively. Density sorting is also required and this in turn needs the presence of at least a weak gravity field. Some meteorite parent bodies must have experienced degassing in their early stage to turn CI compositions into ordinary chondrite compositions and may have had thick dusty surfaces that were easily mobilized by gases evolving from the interior. Density and size sorting may have occurred in the surface layers as the upward drag forces of gases (mainly water) acted against the downward force of gravity. This process is readily modeled quantitatively because it is analogous to the industrially important process of fluidization (Huang *et al.*, 1996b). From fluid dynamics in porous media we calculate gas flow velocities and gas fluxes for the regolith of an asteroid-sized object heated by the impact of accreting objects and by ^{26}Al, and we find that both provide sufficient gas velocities and fluxes for fluidization (Table 7.1). The size and density sorting expected during this process can quantitatively explain metal and chondrule size sorting and distribution in ordinary chondrites. This scenario is broadly in agreement with the major properties of chondritic meteorites (i.e. redox state, petrologic type, cooling rate, matrix abundance, lithophile elemental ratio, etc.).

In summary, I think that most of the properties of chondrites are a result of processes occurring on the surfaces of asteroids, in a thick dynamic regolith or megaregolith. Chondrules are crystallized impact-melt spherules, whose bulk composition and redox state depends on the severity of the impact process. Thus the redox state of the chondrite was inherited largely from the chondrule-forming process, as suggested nearly forty years ago by Larimer and Anders (1967). Heating of a water-rich CI-like asteroid, possibly by internal radioactivities, probably by impact, produced large amounts of gaseous water that sorted the components, created the metal–silicate fractionation, and removed fine-grained matrix by lifting grains clear of the surface in much the same way as dust is removed from comets (Sears et al., 1998). Kitamura and Tsuchiyama (1996) reproduced many of the physical properties of chondrites by shocking porous targets and suggested that the porosity was due to the presence of water.

So what image should we carry in our minds for the formation of chondrites and chondrules, these most primitive materials in the Solar System? The petrographers of the nineteenth century looked at their textures and thought they were volcanic, formed in a plume of gas, dust, and fragments somewhat analogous to the plume of Mount Saint Helens (Fig. 8.1). Volcanism on the scale of Mount Saint Helens can take many forms depending on the composition of the magma and the amount of volatiles, and sometimes the components of the volcanic dust can be carried large distances by hot gases. During this process the components can become fluidized (Wilson, 1980) and size sorted (Fig. 8.2).

But chondrites are clearly not igneous rocks. Despite having many textural features of volcanic breccias, they do not have the compositions or the mineralogy of igneous rocks. They appear to be a non-igneous equivalent of volcanic breccias, composed of material with cosmic composition. They came from a dust-rich environment in which Fe-rich and Fe-poor chondrules could form, volatiles could be lost and sometimes recaptured, and metal and silicates could become separated. They come from an environment in which chondrules still in the plastic state could occasionally collide and undergo multiple recycling. Dodd (1978a; 1978b) described in some detail the complex history of a particular chondrule (Fig. 8.3).

This is essentially the scenario I favor for the formation of chondrules and chondrites. A massive impact on an asteroid with essentially the composition of a CI chondrite, within a few million years of the formation of the oldest solids in the Solar System, caused an event not unlike the explosion of a large volcano. Melt droplets, dust, fragments, and gas were sent up into a plume that engulfed the asteroid and gradually settled down onto its surface. This was the environment in which

8.1 Chondrules and chondrite classes as impact pyroclastics

Figure 8.1 The spectacular explosive eruption of Mount Saint Helens on July 22, 1980, during which pumice and ash were ejected 10–18 km into the air, was visible from Seattle, Washington, 160 km to the north. The view here is from the south. (USGS photograph by Michael P. Doukas.)

chondrules formed, volatiles were lost and occasionally recondensed, oxygen isotopes were fractionated, and metal and silicates were separated. There were to be future impacts and regolith working, enough to stir the surface, cause more brecciation, but probably not enough to produce more chondrules. Chondrites are volcanic breccias, as the early petrologists suggested, but the source of the volcanism was not internally generated heat, but massive impacts; impacts of the scale we have now seen on asteroids, in which the radius of the transient crater is comparable to the diameter of the target asteroid. The nature of the surface remaining after this

154 *How far have we come and where do we go next?*

Figure 8.2 Pyroclastic flow deposit on the Merapi Volcano, Java, Indonesia. Collapse of the lava dome in November 1994 generated pyroclastic flows and surges that traveled as far as 7.5 km from the summit. Over 6000 people were evacuated but despite this the flows and surges killed 43 people. The plume associated with the dome collapse rose 10 km above the volcano. Notice that components are size sorted in these flows, and the overall similarity in the texture between these flows and that of meteorite breccias (Figs. 1.11 and 2.4). (USGS photograph, November 2, 1982, by Jack Lockwood.)

event depended on the intensity of the event but in all probability left the interior unchanged, and of very low density (Fig. 8.4).

8.2 The details

8.2.1 The precursor material

The primary material from which chondrules and chondrites formed was interstellar dust (Clayton, 1983) and trace amounts of grains thought to be presolar have beed found in meteorites (Anders, 1988; Nuth, 1988; Zinner, 1988). While planets were forming at stable locations throughout the Solar System, gravitational instabilities were leading to the production of asteroids in the Asteroid Belt and comets were forming in the outer regions of the Solar System (Cameron, 1995). It is difficult to distinguish between asteroids and comet nuclei, the main difference being formation location, but the orbits of both can become highly eccentric and some fraction of

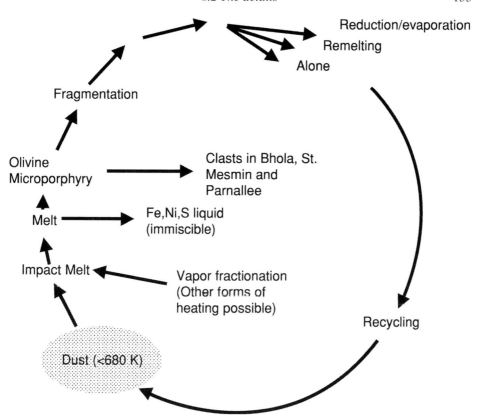

Figure 8.3 Diagram (after Dodd, (1978a; 1978b)) describing a possible evolution sequence for chondrules in ordinary chondrites. Dust is assumed to have formed below about 680 K, so that sulfides are a stable component of the solids. The dust is impact-melted to form metal/sulfide-rich and olivine-rich immiscible melts, perhaps with vapor loss, which are then later fragmented to form clastic chondrules (perhaps with some abrasion into spherules), while others are melted by subsequent impact with reduction and further evaporation. All these materials can be recycled multiple times.

the asteroids and the comets became Earth-crossing. Thus we expect the precursor material to resemble the nucleus of fresh comets. It is often suggested that this would be CI-like, but in view of the rigors of reaching Earth – especially passage through the atmosphere – primitive material might be more carbonaceous and more water-rich than CI chondrites.

8.2.2 The carbonaceous chondrites

Carbonaceous chondrites are the least-altered meteorites and, of the meteorites reaching the Earth's surface, best represent the original material that accreted

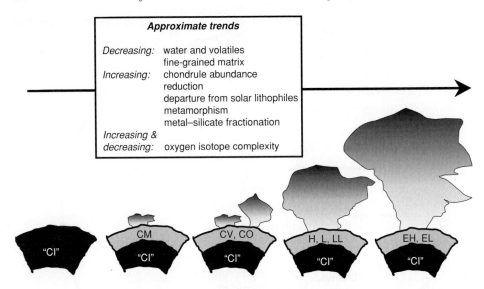

Figure 8.4 The mechanism favored by the author for the origin of chondrules and chondrites. Chondrules and chondrites are the result of major impact (maybe repeated impact) onto a volatile-rich parent body much like the larger C asteroids. Chondrules and refractory inclusions are impact-melt objects that fell back to the surface along with matrix, metal, sulfide, and other chondritic components that settled through a cloud of dust and vapor during which size and density sorting occurred. Major impact onto a primitive (CI or comet-nucleus-like, hence the quotes around "CI") asteroid would more closely resemble pyroclastic volcanism than an impact on the dry Moon. In addition to aerodynamic sorting in the temporary atmosphere, long flight times and slow cooling would result from the almost weightless conditions. To a first approximation the intensity of the impact determines the nature of the fall-back deposit (and class of the chondrite) although variations in projectile size and composition, heterogeneity in the target, heterogeneity in the fall-back deposits, and timing of the event will affect the nature of the deposits. The fall-back debris may well be additionally altered by residual fluids and impact. The trends indicated refer to class average properties; metamorphism, for example, will be less intense for material near the fringes of the deposits. Oxygen isotope properties are particularly complex being determined by impact intensity, timing, and subsequent reactions, but on average result in increased diversity and movement to the lighter end of the terrestrial line and then homogeneity around the center of the presently observed distribution.

together in the early Solar System. In this sense they are the most primitive meteorites. What the carbonaceous chondrites tell us is that the least-altered Solar System materials have no chondrules and no metal and that they are water-rich. They contain up to $\sim$20 vol.% water and $\sim$4 wt % carbon, but these figures are lower limits to that of truly primitive material in the Asteroid Belt.

The chondrites most closely related to the CI chondrites are the CM chondrites which contain only about 10 vol.% water $\sim$10 vol.% of chondrules. The

CM chondrites are therefore meteorites that have experienced some processes that caused a small loss of volatiles and addition of some chondrules.

The trend of decreasing water and increasing chondrule abundance continues with the CV and CO chondrites that are ~35 vol.% chondrule and contain no water. Thus even among the carbonaceous chondrites there is an indication that the formation of chondrules was somehow associated with the loss of water from the meteorites. These chondrite classes begin to show signs of reduction, some of the CV chondrites are much more reduced than others, and they show signs of significant metamorphism. There is also a very small metal–silicate fractionation separating the CV from CO chondrites, and size sorting of chondrules occurred because CO chondrules are very small while CV chondrules are very large. The CV asteroid experienced an early major impact that resulted in the formation of the refractory inclusions abundant in this class. ^{16}O was fractionated by an unidentified physical mechanism during this process (although several mechanisms have been proposed) and the newly formed inclusions reacted with water and other gases in the impact clouds to produce the CCAM line and the FUN inclusions.

8.2.3 The ordinary chondrites

The three classes of ordinary chondrite are very similar in containing essentially no water but ~75 vol.% chondrules. They differ from each other mostly in sizes of their chondrules and metal grains and in the amount of metal and oxidation state. In other words, they all represent meteorites that have lost their original water – assuming that they originally resembled CI chondrites – and whose components have been size sorted during settling in the impact cloud.

Their chondrules underwent similar fractionation in ^{16}O but the details of the process made these chondrules ^{16}O-poor but this isotope reentered the chondrules by subsequent reaction to produce the OCL line. This process was described earlier (see Fig. 4.14). Reactions continued on a much smaller scale during subsequent metamorphism, for instance residual carbon reacted with silicates to lose O as CO or CO_2 and move the meteorites along the terrestrial fraction line from unequilibrated to equilibrated chondrites. In, Tl, and Bi may have been lost and recondensed during settling of the fall-back cloud (so that the accretion temperatures of Larimer and Anders may represent temperatures during the settling phase) but they might also have been lost during auto or prograde metamorphism in the fall-back deposits.

8.2.4 The enstatite chondrites

The enstatite chondrites have a history somewhat similar to the ordinary chondrites, their main distinguishing feature being their high state of reduction. How they came to be highly reduced is unclear. The amount of reducing agent required is actually fairly minimal, a change in C/Si by less than a factor of two is all that is required (Larimer and Bartholomay, 1979). It might be that primitive asteroid material has a C/Si content a factor of two higher than that of CI chondrites. Alternatively, a small random increase in the amount of carbon in the preimpact regolith might be responsible.

The enstatite chondrites suffered severe metal–silicate fractionation and they suffered considerable size sorting of chondrules and metal during the settling of the impact debris cloud to produce the EH and EL chondrules. Following the impact, the two enstatite chondrite classes had very different thermal histories. The EL chondrites were buried and suffered considerable metamorphism which resulted in slow post-metamorphic cooling rates and an almost complete destruction of chondrules. On the other hand, EH chondrites seem to have cooled rapidly and escaped intense metamorphism. These different processes are consistent with different locations in the impact debris cloud.

It seems that the intensity of the impact experienced by the enstatite chondrites was the most severe of all the classes, even resulting in some evaporative loss of Mg relative to the other lithophiles. Oxygen isotopes were completely homogenized by this process, ending up with values close to mean Solar System values near the interception of the CCAM, OCL, and TFL lines.

8.2.5 The metal-rich chondrites

The metal-rich chondrites, the CH chondrites, are meteorites that do not readily fit most models and expected trends. They are clearly chondrites, but they contain very large amounts of metal. They are not impact melts and do not appear to have had the metal introduced by impact events, although there are anomalous meteorites for which this might be the case. The CH chondrites clearly represent a region on a chondritic parent body in which there was an enrichment in metal particles near the surface layers of the postulated regolith described here. In principle, any of the classes should have regions of metal-rich material, but whether it is sampled in the Earth's meteorite collections depends on the vagaries of impact and orbital mechanics.

8.3 So far, so near

While I think both chondrules and their hosts, the chondrites, are impact pyroclastics, formed by impacts in the asteroid parent body, it should be stressed that there is no consensus in the scientific community. In fact, as stated earlier, if a poll were conducted it would probably indicate that most North American researchers favor an origin for chondrules by one of the nebular mechanisms while most European and Japanese researchers favor an impact origin. There is probably equal uncertainty over where or how the volatile element patterns in chondrites were produced, an equal proportion of researchers favoring evaporative loss during chondrule formation and those favoring loss by a process occurring in the nebula. Most of the ideas in the literature for the fractionation of silicates and metal also involve processes occurring in the nebula. The bewildering array of theories for explaining the properties of these primitive rocks is an indication of the difficulty in finding satisfactory explanations. It is also an indication of the amount of effort that has been expended in trying to determine solutions.

So, the best-studied rocks in the Solar System are arguably the least well-understood. In the two hundred years in which meteorites have been studied we have come a long way in collecting data and ideas, but definitive answers to how these rocks became the way they are and what they are telling us about the formation and early history of the Solar System – including our own planet – elude us. However, much has changed in Solar System exploration in the last few years. Solutions to some of the fundamental questions surrounding the oldest rocks in the Solar System might actually be just around the corner.

8.4 Why the impasse?

When the amount of data is large and theories are so numerous, it might reasonably be argued that something is missing. On the one hand we have the asteroids – clearly fragments of a planet that never was, clearly fragments of primitive Solar System material. Astronomers have detected over 20 000 asteroids and have produced spectra of many hundreds of them. Those spectra show exquisite details that can be interpreted in terms of mineral mixtures making up the surface. Considerable effort has been spent using data for lunar samples to understand how exposure to the space environment could affect those spectra. Spacecraft have flown by several asteroids to give us images and quite reasonable estimates for the major geophysical properties. On the other hand we have an extraordinary array of chemical, physical, and isotopic data for meteorites and the most sophisticated equipment has been applied to their study. Incredibly precise information has been obtained on all manner of their properties and small time intervals have been measured for events that occurred billions of years ago. Yet, as we have seen, definitive answers still elude us. Some of the most fundamental questions remain about the meteorites and how they relate to the asteroids. We can be guided in our understanding of space weathering, but we are haunted by the differences in space environment between the Moon and the asteroids, and between their surface compositions. We do not really know how space exposure affects the surfaces of asteroids. If we assume the effect is minimal, then the common types of meteorite are rare – almost to the point of non-existence – in the Asteroid Belt and asteroids commonly appear to resemble rare or non-existent meteorites. There is a veritable log jam of data, and the flow of understanding has become a trickle.

So why the impasse? I suggest that it is because aside from not really knowing how astronomical spectra for asteroids are affected by exposure to space for millions of years, meteorites are – in effect – cosmic jetsam, rocks floating in space that have become washed up on our terrestrial shores. With jetsam, we lack critical information, like a knowledge of what type of asteroid the meteorite came from and what sort of geology on the asteroid produced the rock we examine in the laboratory. A terrestrial geologist who collected his or her samples from the rocks washed down to stream beds or beaches would not get very far in understanding the region he or she was interested in. It is necessary to travel to the actual outcrop. The geologist must know where the rock formed and how it related to the surrounding strata. This is absolutely crucial information.

8.5 Breaking the log jam?

When we have asteroids tens of millions of miles away and only spectra to look at, and we have rocks that nature provided with no clue to origin, the answer is to go to the asteroid for which we have spectra and bring a sample back to examine in the laboratory using the same techniques honed by meteorite researchers over the last two centuries. This will provide the most direct linkage between the meteorites and the asteroids. At the same time we can collect samples from known sites on the asteroids, from inside craters, from crater walls, from crater ejecta, from intercrater plains, from fine-grained ponds, from unsual linear features or scarps, from regolith, or from bedrock. Comparing rocks from different geological histories will provide unique information on the processes occurring on asteroids and help disentangle processes occurring on the asteroid from those occurring in the nebula. Sample return from asteroids is now recognized as a high priority for Solar System exploration (Sears *et al.*, 2002; Space Studies Board, 2002).

The Japanese space agency's Hayabusa spacecraft (named MUSES C before launch) was launched in May 2003 by ISAS' M-V launch vehicle. After 22 months it is due to arrive at the near-Earth asteroid 1998SF36 (Fig. 8.5). During a three-month stay in the vicinity of the asteroid, the spacecraft will make observations of the surface and then take samples using a collector that fires a projectile into the surface and collects the ejecta. The main engine is a solar electric propulsion thruster that uses sunlight to generate electricity that creates an electric field to derive thrust by propelling Xe ions out of the engine. The mission will conclude in

Figure 8.5 The Hayabusa spacecraft. Hayabusa is an ISAS spacecraft that will return the first samples of a near-Earth asteroid to Earth. The spacecraft will visit an asteroid, descend to the surface, and place a cone on the surface through which a projectile will be fired. Ejecta thrown up from the surface will be funneled into a collector inside a sample return capsule. (Courtesy H. Yano, ISAS.)

the summer of 2007, when the reentry capsule separates from the spacecraft, enters the Earth atmosphere, and is retrieved on the ground. Total mission time is about 4.5 years.

Other near-Earth asteroid sample return missions are being planned (Sears *et al.*, 2002; Franzen and Sears, 2003). For example, the Hera mission is proposed to visit three asteroids and take three samples from each, a total of nine samples with a combined weight of nearly 1 kg (Fig. 8.6). Like Hayabusa it will have solar electric propulsion, but it will carry three ion thrusters firing only two at a time. It will spend at least two months studying the surface of the asteroid before sweeping down to pick up surface samples with an adhesive pad. In this way, the scientists will have time to select the sites for sampling so as to maximize the scientific value of the samples. After about six years in space, Hera will return to Earth with its samples.

No doubt these samples will uncover new facts and prompt new questions, but hopefully they will resolve some of the long-lasting questions that meteorites have not yet answered in two centuries of study. What returned samples from near-Earth asteroids will do might be suggested by our experience with samples from the

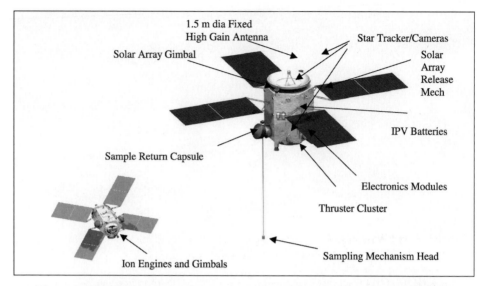

Figure 8.6 The Hera spacecraft. Hera is a proposed NASA mission that will visit multiple near-Earth asteroids, perform a photoreconnaisance and then collect several samples from each using a sticky pad on the end of a boom. The boom will then stack the samples in a container for return to Earth. (Courtesy J. Preble, SpaceWorks, Inc.)

Moon. Returned samples will fundamentally alter many of our notions of what the early Solar System was like, and may turn up some major new ideas. In the words of the NASA policy makers, samples will be "paradigm altering." Samples will do many things, but they will not end our fascination with the oldest rocks in the Solar System.

References

Afiattalab F. and Wasson J. T. (1980) Composition of the metal phases in ordinary chondrites: implications regarding classification and metamorphism. *Geochim. Cosmochim. Acta* **44**, 431–46.

Ahrens L. H. (1965) Observations on the Fe–Si–Mg relationship in chondrites. *Geochim. Cosmochim. Acta* **29**, 801–6.

  (1970) The composition of stony meteorites (VII): Observations on fractionation between the L and H chondrites. *Earth Planet. Sci. Lett.* **9**, 345–7.

Ahrens L. H. and Von Michaelis H. (1969) The composition of stony meteorites III. Some inter-element relationships. *Earth Planet. Sci. Lett.* **5**, 395–400.

Ahrens T. J., O'Keefe J. D. and Lange M. A. (1989) Formation of atmospheres during accretion of the terrestrial planets. In *Origin and Evolution of Planetary and Satellite Atmospheres*. Ed. S. K. Atreya, J. B. Pollack and M. S. Matthews. University of Arizona Press, pp. 328–85.

Akridge G., Benoit P. H. and Sears D. W. G. (1998) Regolith and megaregolith formation of H-chondrites: Thermal constraints on the parent body. *Icarus* **132**, 185–95.

Alexander C. M. O'D. (1994) Trace element distributions within ordinary chondrite chondrules: Implications for chondrule formation conditions and precursors. *Geochim. Cosmochim. Acta* **58**, 3451–67.

  (1996) Recycling and volatile loss in chondrule formation. In *Chondrules and the Protoplanetary Disk*. Ed. R. H. Hewins, R. H. Jones and E. R. D. Scott. Cambridge University Press, pp. 233–41.

Alexander C. M. O'D and Wang J. (2001) Iron isotopes in chondrules: Implications for the role of evaporation during chondrule formation. *Meteorit. Planet. Sci.* **36**, 419–28.

Alexander C. M. O'D., Hutchison R. and Barber D. J. (1989) Origin of chondrule rims and interchondrule matrices in unequilibrated ordinary chondrites. *Earth Planet. Sci. Lett.* **95**, 187–207.

Alexander C. M. O'D., Grossman J. N., Wang J., *et al.* (2000) The lack of potassium-isotopic fractionation in Bishunpur chondrules. *Meteorit. Planet. Sci.* **35**, 859–68.

Allen I. S., Nozette S. and Wilkening L. L. (1980) A study of chondrule rims and chondrule irradiation records in unequilibrated ordinary chondrites. *Geochim. Cosmochim. Acta* **44**, 1161–75.

Anders E. (1964) Origin, age and composition of meteorites. *Space Sci. Rev.* **3**, 583–714.

(1977) Critique of "Nebular condensation of moderately volatile elements and their abundances in ordinary chondrites" by C. M. Wai and J. T. Wasson. *Earth Planet. Sci. Lett.* **36**, 14–20.

(1988) Circumstellar material in meteorites: noble gases, carbon and nitrogen. In *Meteorites and the Early Solar System*. Ed. J. F. Kerridge and M. S. Matthews. University of Arizona Press, pp. 927–55.

Anders E., Higuchi H., Ganapathy R. and Morgan J. W. (1976) Chemical fractionations in meteorites – X. C3 chondrites. *Geochim. Cosmochim. Acta* **40**, 1131–9.

Ashworth J. R. (1977) Matrix textures in unequilibrated ordinary chondrites. *Earth Planet. Sci. Lett.* **35**, 25–34.

Asphaug E. and Nolan M. C. (1992) Analytical and numerical predictions for regolith thickness on asteroids (abstract). *Lunar Planet. Sci.* **XXIII**, 43–4.

Baldwin B. and Shaeffer Y. (1971) Ablation and breakup of large meteoroids during atmospheric entry. *J. Geophys. Res.* **76**, 4653–68.

Ball R. S. (1910) *The Story of the Heavens*. Cassell and Co.

Batchelor J. D., Symes S. J. K., Benoit P. H. and Sears D. W. G. (1997) Constraints on the thermal and mixing history of lunar surface materials and comparisons with basaltic meteorites. *J. Geophys. Res.* **102**, 19321–35

Bell J. F. and Keil K. (1988) Spectral alteration effects in chondritic gas-rich breccias: Implication for S-class and Q-class asteroids. *Proc. 18th Lunar Planet. Sci. Conf.* Lunar and Planetary Institute, pp. 573–80.

Bell J. F., Davis D. R., Hartmann W. K. and Gaffey M. J. (1989) Asteroids: The big picture. In *Asteroids II*. Ed. R. P. Binzel, T. Gehrels and M. S. Matthews. University of Arizona Press, pp. 921–45.

Belton M. J. S., Veverka J., Thomas P., *et al.* (1992) Galileo encounter with 951 Gaspra – First pictures of an asteroid. *Science* **257**, 1647–52.

Bennett M. E. and McSween H. Y. (1996) Revised model calculations for the thermal histories for ordinary chondrite parent bodies. *Meteorit. Planet. Sci.* **31**, 783–92.

Benoit P. H. and Sears D. W. G. (1992) The breakup of a meteorite parent body and the delivery of meteorites to Earth. *Science* **255**, 1685–7.

(1993) A recent meteorite shower in Antarctica with an unusual orbital history. *Earth Planet. Sci. Lett.* **120**, 463–71.

(1996) Rapid changes in the nature of the H chondrites falling to Earth. *Meteorit. Planet. Sci.* **31**, 81–6.

Berwerth F. M. (1901) *Centralblatt Min.* **21**, 641–7. (Cited in Merrill, 1920.)

Berzelius J. J. (1834) Uber Meteorstein. *Annal. Physik* **33**, 1–32, 113–48.

Bhandari N., Goswami J. N., Gupta S. K., *et al.* (1972) Collision controlled radiation history of the lunar regolith. *Proc. 3rd Lunar Sci. Conf.* Lunar and Planetary Institute, pp. 2811–29.

Binzel R. P., Gehrels T. and Matthews M. S., Eds. (1989) *Asteroids II*. University of Arizona Press.

Binzel R. P., Schelte J. B., Burbine T. H. and Sunshine J. M. (1996) Spectral properties of near-Earth asteroids: Evidence for sources of ordinary chondrite meteorites. *Science* **273**, 946–8.

Birck J.-L. and Allegre C. J. (1985) Evidence for the presence of ^{53}Mn in the early solar system. *Geophys. Res. Lett.* **12**, 745–8.

Bischoff A. (1998) Aqueous alteration of carbonaceous chondrites: Evidence for preaccretionary alteration – a review. *Meteorit. Planet. Sci.* **33**, 1113–22.

Bischoff A. and Keil K. (1983) Ca–Al-rich chondrules and inclusions in ordinary chondrites. *Nature* **303**, 588–92.

Bischoff A., Palme H., Weber H. W., *et al.* (1987) Petrography, shock history, chemical composition and noble gas content of the lunar meteorites Yamato-82192 and -82193. *Mem. Natl. Inst. Polar Res., Spec. Issue* **46**, 21–42.

Bischoff A., Palme H., Ash R. D., *et al.* (1993a) Paired Renazzo-type (CR) carbonaceous chondrites from the Sahara. *Geochim. Cosmochim. Acta* **57**, 1587–603.

Bischoff A., Palme H., Schultz L., *et al.* (1993b) ACFER 182 and paired samples, an iron-rich carbonaceous chondrite – Similarities with ALH85085 and relationship to CR chondrites. *Geochim. Cosmochim. Acta* **57**, 2631–48.

Blander M. (1975) Critical comments on a proposed cosmothermometer. *Geochim. Cosmochim. Acta* **39**, 1315–20.

  (1983) Condensation of chondrules. In *Chondrules and Their Origins*. Ed. E. A. King. Lunar and Planetary Institute, pp. 1–9.

Blander M. and Katz J. L. (1967) Condensation of primordial dust. *Geochim. Cosmochim. Acta* **31**, 1025–34.

Bogard D. D. (1994) Impact ages of meteorites: A synthesis. *Meteoritics* **30**, 244–68.

  (1995) ^{39}Ar–^{40}Ar ages of two shocked L chondrites (abstract). *Lunar Planet. Sci.* **XXVI**, 141–2.

Borg J., Chaumont J., Jouret C., Langevin Y. and Maurette M. (1980) Solar wind radiation damage in lunar dust grains and the characteristics of the ancient solar wind. In *Proceedings of a Conference on the Ancient Sun*. Ed. R. O. Pepin, J. A. Eddy and R. B. Merrill. Pergamon, pp. 431–61.

Borgstrom L. H. (1904) The Shelburne meteorite. *Trans. Roy. Astron. Soc. Canada* pp. 69–94.

Boss A. P. (1988) High temperatures in the early solar nebula. *Science* **241**, 565–7.

  (1993) Evolution of the solar nebula. II. Thermal structure during nebula formation. *Astrophys. J.* **417**, 351–67.

  (1996a) Large scale processes in the solar nebula. In *Chondrules and the Protoplanetary Disk*. Ed. R. H. Hewins, R. H. Jones and E. R. D. Scott. Cambridge University Press, pp. 29–34.

  (1996b) A concise guide to chondrule formation models. In *Chondrules and the Protoplanetary Disk*. Ed. R. H. Hewins, R. H. Jones and E. R. D. Scott. Cambridge University Press, pp. 257–63.

Boss A. P. and Graham J. A. (1993) Clumpy disk accretion and chondrule formation. *Icarus* **106**, 168–78.

Boynton W. V. (1975) Fractionation in the solar nebula – Condensation of yttrium and the rare earth elements. *Geochim. Cosmochim. Acta* **39**, 569–84.

Bradley J. P., Sandford S. A. and Walker R. M. (1988) Interplanetary dust particles. In *Meteorites and the Early Solar System*. Ed. J. F. Kerridge and M. S. Matthews. University of Arizona Press, pp. 861–95.

Brearley A. J. (1993) Matrix and fine-grained rims in the unequilibrated CO3 chondrite, ALHA77307 – Origins and evidence for diverse, primitive nebular dust components. *Geochim. Cosmochim. Acta* **57**, 1521–50.

  (1996) Nature of matrix in unequilibrated chondrites and its possible relationship to chondrules. In *Chondrules and the Protoplanetary Disk*. Ed. R. H. Hewins, R. H. Jones and E. R. D. Scott. Cambridge University Press, pp. 137–51.

Brearley A. J. and Geiger T. (1991) Mineralogical and chemical studies bearing on the origin of accretionary rims in the Murchison CM2 carbonaceous chondrite. *Meteoritics* **26**, 323.

Brearley A. J., Scott E. R. D., Keil K., *et al.* (1989) Chemical, isotopic and mineralogical evidence for the origin of matrix in ordinary chondrites. *Geochim. Cosmochim. Acta* **53**, 2081–93.

Brezina A. (1885) *Die Meteoritiensammlung des k. k. mineralogischen Hofkabinetes in Wein am 1 Mai 1885*. Alfred Hölder.

Bridges J. C. (1999). Mineralogical controls on the oxygen isotopic compositions of UOCs. *Geochim. Cosmochim. Acta* **63**, 945–51.

Bridges J. C., Franchi I. A., Hutchsion R., Sexton A. S. and Pillinger C. T. (1997) Mineralogical and oxygen isotopic constraints on the formation of Chainpur (LL3) and Parnallee (LL3) chondrules (abstract). *Lunar Planet. Sci.* **XXVIII**, 155–6.

Bridges J. C., Franchi I. A., Hutchsion R., Sexton A. S. and Pillinger C. T. (1998) Correlated mineralogy, chemical compositions, oxygen isotopic composition and sizes of chondrules. *Earth Planet. Sci. Lett.* **155**, 183–96.

Brigham C. A., Yabuki H., Ouyang Z., *et al.* (1986) Silica-bearing chondrules and clasts in ordinary chondrites. *Geochim. Cosmochim. Acta* **50**, 1655–66.

Britt D. T. and Consolmagno G. (2002) Stony meteorite porosities and densities: A review of data through 2001. Unpublished paper.

Britt D. T. and Pieters C. M. (1991) Darkening in gas-rich ordinary chondrites: Spectral modeling and implications for the regoliths of ordinary chondrite parent bodies (abstract). *Lunar Planet. Sci.* **XXII**, 141–2.

Britt D. T., Yeomans D., Housen K. and Consolmagno G. (2002) Asteroid density, porosity and structure. In *Asteroids III*. Ed. W. F. Bottke, A. Cellino, P. Paolicchi and R. P. Binzel. University of Arizona Press, pp. 485–500.

Browning L. B., McSween H. Y. and Zolensky M. E. (1996) Correlated alteration effects in CM carbonaceous chondrites. *Geochim. Cosmochim. Acta* **60**, 2621–33.

Brownlee D. E. and Rajan R. S. (1973) Micrometeorite craters discovered on chondrule-like objects from the Kapoeta meteorite. *Science* **182**, 1341–4.

Brownlee D. E., Bates B. and Beauchamp R. H. (1983) Meteor ablation spherules as chondrule analogs. In *Chondrules and Their Origins*. Ed. E. A. King. Lunar and Planetary Institute, pp. 10–25.

Brownlee D. E., Bates B. and Schramm L. (1997) The elemental composition of stony cosmic spherules. *Meteoritics* **32**, 157–75

Buchwald V. F. (1975) *Handbook of Iron Meteorites, Their History, Composition and Structure*. University of California Press.

(1992) *Meteoritter – nøglen til Jordens fortid*. GlydendaL.

Bunch T. E. and Chang S. (1980) Carbonaceous chondrites–II. Carbonaceous chondrite phyllosilicates and light element geochemistry as indicators of parent body processes and surface conditions. *Geochim. Cosmochim. Acta* **44**, 1543–77.

Bunch T. E. and Rajan R. S. (1988) Meteorite regolithic breccias. In *Meteorites and the Early Solar System*. Ed. J. F. Kerridge and M. S. Matthews. University of Arizona Press, pp. 144–64.

Bunch T. E., Quaide W., Prinz M., Keil K. and Dowty E. (1972) Lunar ultramafic glasses, chondrules and rocks. *Nat. Phys. Sci.* **239**, 57–9.

Bunch T. E., Chang S., Cassen P., Reynolds R. and Lissauer J. (1985) Non-nebula origin for CAI rims (abstract). *Lunar Planet. Sci.* **XVI**, 97–8.

Burbage E. M., Burbage G. R., Fowler W. A. and Hoyle F. (1957) Synthesis of the elements in stars. *Rev. Mod. Phys.* **29**, 547–640.

Burbine T. H., McCoy T. J., Meibom A., Gladman B. and Keil K. (2003) Meteorite parent bodies: Their number and identification. In *Asteroids III*. Ed. W. F. Bottke, A. Cellino, P. Paolicchi and R. P. Binzel. University of Arizona Press, pp. 653–67.

Burke J. G. (1986) *Cosmic debris: Meteorites in History*. University of California Press.

Butler R. F. (1972) Natural remanent magnetization and thermomagnetic properties of the Allende meteorite. *Earth Planet. Sci. Lett.* **17**, 120–8.

Caffee M. W. and Macdougall J. D. (1988) Compaction ages. In *Meteorites and the Early Solar System*. Ed. J. F. Kerridge and M. S. Matthews. University of Arizona Press, pp. 289–98.

Caffee M. W., Hohenberg C. M., Swindle T. D. and Goswami J. N. (1987) Evidence in meteorites for an active early Sun. *Astrophys. J.* **313**, L31–5.

Caffee M. W., Goswami J. N., Hohenberg C. M., Marti K. and Reedy R. C. (1988) Irradiation records in meteorites. In *Meteorites and the Early Solar System*. Ed. J. F. Kerridge and M. S. Matthews. University of Arizona Press, pp. 205–45.

Cameron A. G. W. (1966) The accumulation of chondritic material. *Earth Planet. Sci. Lett.* **1**, 93–6.

  (1995) The first ten million years in the solar nebula. *Meteoritics* **30**, 133–61.

Cameron A. G. W. and Fegley M. B. (1982) Nucleation and condensation in the primitive solar nebula. *Icarus* **52**, 1–13.

Carr M. H., Kirk R., McEwen A., *et al.* (1994) The geology of Gaspra. *Icarus* **107**, 61–71.

Cassen P. (1994) Utilitarian models of the solar nebula. *Icarus* **112**, 405–29.

  (1996a) Overview of models of the solar nebula: Potential chondrule-forming environments. In *Chondrules and the Protoplanetary Disk*. Ed. R. H. Hewins, R. H. Jones and E. R. D. Scott. Cambridge University Press, pp. 21–8.

  (1996b) Models for the fractionation of moderately volatile elements in the solar nebula. *Meteoritics* **31**, 793–806.

  (2001) Nebular thermal evolution and the properties of primitive planetary materials. *Meteorit. Planet. Sci.* **36**, 671–700.

Cassen P. and Boss A. P. (1988) Protostellar collapse, dust grains and solar system formation. In *Meteorites and the Early Solar System*. Ed. J. F. Kerridge and M. S. Matthews. University of Arizona Press, pp. 304–28.

Castaing R. (1952) Application des sondes electronique a une methode d'analyse ponctuelle chimique et crystallographique. *Office Nat. d'Études Res. Aéronaut.* **55**, 27–31.

Chambers J. E. and Cassen P. (2002) The effects of nebula surface density profile and giant-planet eccentricities on planetary accretion in the inner solar system. *Meteorit. Planet. Sci.* **37**, 1523–40.

Chambers J. E. and Wetherill G. W. (1998) Making the terrestrial planets: N-body integrations of planetary embryos in three dimensions. *Icarus* **136**, 304–27.

Chapman C. R. (1976) Asteroids as meteorite parent-bodies: The astronomical perspective. *Geochim. Cosmochim. Acta* **40**, 701–19.

  (1996) S-type asteroids, ordinary chondrites, and space weathering: The evidence from Galileo's fly-bys of Gaspra and Ida. *Meteoritics* **31**, 699–725.

  (2001) Eros at very high resolution: Meteoritical implications. *Meteorit. Planet. Sci.* **36**, Supplement, A39.

Chapman C. R., Veverka J., Thomas P. C., *et al.* (1995) Discovery and physical properties of Dactyl A satellite of asteroid 243 Ida. *Nature* **374**, 783–5.

Chapman C. R., Ryan E. V., Merline W. J., *et al.* (1996a) Cratering on Ida. *Icarus* **120**, 77–86.

Chapman C. R., Veverka J., Belton M. J. S., Neukum G. and Morrison D. (1996b) Cratering on Gaspra. *Icarus* **120**, 231–45.

Chapman C. R., Merline W. J. and Thomas P. (1999) Cratering on Mathilde. *Icarus* **140**, 28–33.

Chapman C. R., Merline W. J., Thomas P. C., *et al.* (2002) Impact history of Eros: craters and boulders. *Icarus* **155**, 104–18.

Chladni E. F. F. (1794) *Ueber den Orsprung der von Pallas gefunden und anderer ihr ähnlicher Eisenmassen.* J. F. Hartknoch.

Christophe-Michel-Lévy M. (1976) La matrice noire et blanche de la chondrite de Tieschitz (H3). *Earth Planet. Sci. Lett.* **30**, 143–50.

(1981) Some clues to the history of H-group chondrites. *Earth Planet. Sci. Lett.* **54**, 67–80.

(1987) Microchondrules in the Mezö-Madaras and Krymka unequilibrated chondrites (abstract). *Meteoritics* **22**, 355–6.

Cirlin E.-H., Taylor L. A. and Lofgren G. E. (1985) Fe/Mg K_D for olivine/liquid in chondrules: Effects of cooling rate (abstract). *Lunar Planet. Sci.* **XVI**, 133–4.

Clarke C. L., Lin D. N. C. and Pringle J. E. (1990) Pre-conditions for discgenerated FU Orionis outbursts. *Mon. Nat. Roy. Astron. Soc.* **242**, 439–46.

Clayton D. D. (1980a) Chemical and isotopic fractionation by grain size separation. *Earth Planet. Sci. Lett.* **47**, 199–210.

(1980b) Chemical energy in cold-cloud aggregates: The origin of meteoritic chondrules. *Astrophys. J.* **239**, L37–41.

(1983) Chemical state of pre-solar matter. In *Chondrules and Their Origins*. Ed. E. A. King. Lunar and Planetary Institute, pp. 26–36.

(1988) Stellar nucleosynthesis and chemical evolution of the solar neighborhood. In *Meteorites and the Early Solar System*. Ed. J. F. Kerridge and M. S. Matthews. University of Arizona Press, pp. 1021–62.

Clayton R. N. (1993) Oxygen isotopes in meteorites. *Ann. Rev. Earth Planet. Sci.* **21**, 115–49.

Clayton R. N. and Mayeda T. K. (1984) The oxygen isotope record in Murchison and other carbonaceous chondrites. *Earth Planet. Sci. Lett.* **67**, 151–61.

(1985) Oxygen isotopes in chondrules from enstatite chondrites: Possible identification of a major nebular reservoir (abstract). *Lunar Planet. Sci.* **XVI**, 142–3.

(1999) Links among CI and CM chondrites. *Lunar Planet. Sci.* **XXX** , abstract no. 1795.

Clayton R. N., Mayeda T. K., Gooding J. L., Keil K. and Olsen E. J. (1981) Redox processes in chondrules and chondrites (abstract). *Lunar Planet. Sci.* **XII**, 154–6.

Clayton R. N., Onuma N., Ikeda Y., *et al.* (1983) Oxygen isotopic compositions of chondrules in Allende and ordinary chondrites. In *Chondrules and Their Origins*. Ed. E. A. King. Lunar and Planetary Institute, pp. 37–43.

Clayton R. N., Mayeda T. K. and Molini-Velsko C. A. (1985) Isotopic variations in solar system material – Evaporation and condensation of silicates. In *Protostars and Planets II*. Ed. D. C. Black and M. S. Matthews. University of Arizona Press, pp. 755–71.

Clayton R. N., Mayeda T. K., Rubin A. E. and Wasson J. T. (1987) Oxygen isotopes in Allende chondrules and coarse-grained rims (abstract). *Lunar Planet. Sci.* **XVIII**, 187–8.

Clayton R. N., Mayeda T. K., Goswami J. N. and Olsen E. J. (1991) Oxygen isotopes studies of ordinary chondrites. *Geochim. Cosmochim. Acta* **55**, 2317–37.

Clayton R. N., Mayeda T. K. and Nagahara H. (1992) Oxygen isotope relationship among primitive achondrites (abstract). *Lunar Planet. Sci.* **XXIII**, 231–2.

Colson R. O., Taylor L. A. and McKay G. A. (1986) Predictive thermodynamic modeling for trace element partitioning in magmatic systems (abstract.). *Lunar Planet. Sci.* **XVII**, 144–5.

Colson R. O., McKay G. A. and Taylor L. A. (1988) Temperature and composition dependencies of trace element partitioning: Olivine/melt and low-Ca pyroxene/melt. *Geochim. Cosmochim. Acta* **52**, 539–53.

Connolly H. C. Jr and Hewins R. H. (1996) Constraints on chondrule precursors from experimental data. In *Chondrules and the Protoplanetary Disk*. Ed. R. H. Hewins, R. H. Jones and E. R. D. Scott. Cambridge University Press, pp. 129–35.

Connolly H. C. Jr, Radomsky P. M. and Hewins R. H. (1988) Chondrule texture: The influence of bulk composition and heating time for uniform thermal conditions (abstract). *Lunar Planet. Sci.* **XIX**, 205–6.

Connolly H. C. Jr, Hewins R. H. and Lofgren G. E. (1993) Flash melting of chondrule precursors in excess of 1600 °C. Series I: Type II (B1) chondrule composition experiments (abstract). *Lunar Planet. Sci.* **XXIV**, 329–30.

Craig H. (1964) Petrological and compositional relationships in meteorites. In *Isotopic and Cosmic Chemistry*. Ed. H. Craig, S. L. Miller and G. J. Wasserburg. North-Holland, pp. 401–51.

Cronin J. R., Pizzarello S. and Cruikshank D. P. (1988) Organic matter in carbonaceous chondrites, planetary satellites, asteroids and comets. In *Meteorites and the Early Solar System*. Ed. J. F. Kerridge and M. S. Matthews. University of Arizona Press, pp. 819–57.

Cuzzi J. N., Dobrovolskis A. R. and Champney J. M. (1993) Particle–gas dynamics in the midplane of a protoplanetary nebula. *Icarus* **106**, 102–34.

Cuzzi J. N., Dobrovolskis A. R. and Hogan R. C. (1996) Turbulence, chondrules, and planetesimals. In *Chondrules and the Protoplanetary Disk*. Ed. R. H. Hewins, R. H. Jones and E. R. D. Scott. Cambridge University Press, pp. 35–43.

Daubree G. A. (1879) *Études Synthétiques de Geologic Expérimentale*. Dunod, p. 530.

Davis A. M. and MacPherson G. J. (1988) Further isotopic and chemical investigations of an isotopically heterogeneous Vigarano inclusion. *Meteoritics* **23**, 264–5.
  (1996) Thermal processing in the solar nebula: Constraints from refractory inclusions. In *Chondrules and the Protoplanetary Disk*. Ed. R. H. Hewins, R. H. Jones and E. R. D. Scott. Cambridge University Press, pp. 71–6.

Davis A. M., Hashimoto A., Clayton R. N. and Mayeda T. K. (1990) Isotope mass fractionation during evaporation of Mg_2SiO_4. *Nature* **347**, 655–8.

Davis D. R., Chapman C. R., Greenberg R., Weidenschilling S. and Harris A. W. (1979) Collisional evolution of asteroids: Populations, rotations, and velocities. In *Asteroids*. Ed. T. Gehrels. University of Arizona Press, pp. 528–57.

DeHart J. M., Lofgren G. E., Lu J., Benoit P. H. and Sears D. W. G. (1992) Chemical and physical studies of chondrites X: Cathodoluminescence studies of metamorphism and nebular processes in type 3 ordinary chondrites. *Geochim. Cosmochim. Acta* **56**, 3791–807.

Delano J. W. (1986) Pristine lunar glasses: Criteria, data and implications. *Proc. 16th Lunar Planet. Sci. Conf., part 2. J. Geophys. Res.* **91**, D201–13.
  (1991) Geochemical comparison of impact glasses from lunar meteorites ALHA81004 and MAC 88105 and Apollo 16 regolith 64001. *Geochim. Cosmochim. Acta* **55**, 3019–29.

Dence M. R. and Plant A. G. (1972) Analysis of Fra Mauro and the origin of the Imbrium Basin. *Proc. 3rd Lunar Sci. Conf.* Lunar and Planetary Institute, pp. 379–99.

Desch S. J. and Connolly H. C. Jr (2002) A model of the thermal processing of particles in solar nebula shocks: Application to the cooling rates of chondrules. *Meteorit. Planet. Sci.* **37**, 183–207.

Dodd R. T. (1967) Particle sizes in and composition of unequilibrated ordinary chondrites (abstract). *Trans. AGU* **48**, 159.
  (1969) Metamorphism of ordinary chondrites: a review. *Geochim. Cosmochim. Acta* **33**, 161–203.

(1971) The petrology of chondrules in the Sharps meteorite. *Contrib. Mineral. Petrol.* **31**, 201–27.

(1973) Minor element abundances in olivines in the Sharps (H-3) chondrite. *Contrib. Mineral. Petrol.* **42**, 159–67.

(1974) The petrology of chondrules in the Hallingeberg meteorite. *Contrib. Mineral. Petrol.* **47**, 97–112.

(1976) Accretion of the ordinary chondrites. *Earth Planet. Sci. Lett.* **28**, 479–84.

(1978a) The composition and origin of large microporphyritic chondrules in the Manych (L-3) chondrite. *Earth Planet. Sci. Lett.* **39**, 52–66.

(1978b) Compositions of droplet chondrules in the Manych (L-3) chondrite and the origin of chondrules. *Earth Planet. Sci. Lett.* **40**, 71–82.

(1981) *Meteorites: A Petrologic–Chemical Synthesis.* Cambridge University Press.

(1986) *Thunderstones and Shooting Stars.* Cambridge University Press.

Dodd R. T. and Teleky L. S. (1967) Preferred orientation of olivine crystals in porphyritic chondrules. *Icarus* **6**, 407–16.

Dodd R. T. and Van Schmus W. R. (1971) Dark-zoned chondrules. *Chem. Erde* **30**, 59–69.

Dodd R. T. and Walter L. S. (1972) Chemical constraints on the origin of chondrules in ordinary chondrites. In *On the Origin of the Solar System*. Ed. H. Reeves. Centre National de la Recherche Scientifique, pp. 293–300.

Dodd R. T., Van Schmus W. R. and Koffman D. M. (1967) A survey of the unequilibrated ordinary chondrites. *Geochim. Cosmochim. Acta* **31**, 921–51.

Donaldson C. H. (1979) Composition changes in a basalt melt contained in a wire loop of $Pt_{80}Rh_{20}$: Effects of temperature, time, and oxygen fugacity. *Mineral. Mag.* **43**, 115–19.

Donn B. and Sears D. G. W. (1963) Planets and comets: Role of crystal growth in their formation. *Science* **140**, 1208–11.

Dubmile B., Morrill G. and Sterzik M. (1995) The dust subdisk in the protoplanetary nebula. *Icarus* **114**, 237–46.

Dunn T. (1987) Partitioning of Hf, Lu, Ti and Mn between olivine, clinopyroxene and basaltic liquid. *Contrib. Mineral. Petrol.* **96**, 476–84.

Eisenhour D. and Buseck P. R. (1993) Primordial lighting: evidence preserved in chondrites (abstract). *Lunar Planet. Sci.* **XXIV**, 435–6.

(1995) Radiative heating and the size distribution of pre-chondrule aggregates of dust (abstract). *Lunar Planet. Sci.* **XXVI**, 365–6.

Eisenhour D. D., Daulton T. L. and Buseck P. R. (1994) Electromagnetic heating in the early solar nebula and the formation of chondrules. *Science* **265**, 1067–70.

Evensen N. M., Carter S. R., Hamilton P. J., O'Nions R. K. and Ridley W. I. (1979) A combined chemical–petrological study of separated chondrules from the Richardton meteorite. *Earth Planet. Sci. Lett.* **42**, 223–36.

Farinella P., Paollicchi P. and Zappalà V. (1982) The asteroids as outcomes of catastrophic collisions. *Icarus* **52**, 409–33.

Farinella P., Vokrouhlicky D. and Hartmann W. K. (1998) Meteorite delivery via Yarkovsky orbital drift. *Icarus* **132**, 378–87.

Farrington O. C. (1905) *Meteorites: Their Structure, Composition and Terrestrial relations.* Published by the author in Chicago.

Fegley B. Jr and Palme H. (1985) Evidence for oxidizing conditions in the solar nebula from Mo and W depletions in refractory inclusions in carbonaceous chondrites. *Earth Planet. Sci. Lett.* **75**, 311–26.

Fermor L. L. (1938) Garnets and their role in nature. *Indian Assoc. Adv. Sci. Spec. Publ.* **6**, 87–91.

Ferraris C., Folco L. and Mellini M. (2002) Chondrule thermal history from unequilibrated H chondrites: A transmission and analytical electron microscopy study. *Meteorit. Planet. Sci.* **37**, 1299–321.

Fisher R. V., Smith A. L. and Roobol M. J. (1980) Destruction of St. Pierre, Martinique, by ash-cloud surges, May 8 and 20, 1902. *Geology* **8**, 472–6.

Fleck R. C. Jr (1990) Comment on 'Magnetic reconnection flares in the protoplanetary nebula and the possible origin of meteorite chondrules'. *Icarus* **87**, 241–3. Reply, 244–6.

Ford C. E., Russell D. G., Craven J. A. and Fisk M. R. (1983) Olivine–liquid equilibria: Temperature, pressure and compositional dependence of the crystal/liquid cation partition coefficients for Mg, Ca and Mn. *J. Petrol.* **24**, 256–65.

Fougeroux A. D., Cadet L. C. and Lavoisier A. (1772) Rapport fait a l'Academie Royale des Science, d'une observation communique par M. L'Abbe Bachelay, sur une Pierre qu'on prètend etre tombèe cu Ciel pendant un orage. Observations sur la physique, sur l'histoire naturelle, et sur les Arts *J. Physique* **2**, 251–5 (printed and dated 1777).

Franzen M. A. and Sears D. W. G. (2003) The Hera near-Earth asteroid sample return mission: An overview (abstract). *Lunar Planet. Sci.* **XXXIV**, abstract no. 1032.

Fredriksson K. (1963) Chondrules and the meteorite parent bodies. *Trans. N. Y. Acad. Sci.* **25**, 756–69.

(1983) Crystallinity, recrystallization, equilibration, and metamorphism in chondrites. In *Chondrules and Their Origins*. Ed. E. A. King. Lunar and Planetary Institute, pp. 44–52.

Fredriksson K. and Ringwood A. E. (1963) Origin of meteoritic chondrules. *Geochim. Cosmochim. Acta* **27**, 639–41.

Fredriksson K., Nelen J. and Fredriksson B. J. (1968) The LL-group chondrites. In *Origin and Distribution of the Elements*. Ed. L. H. Ahrens. Pergamon, pp. 457–66.

Fredriksson, K., Jarosewich, E. and Nelen, J. (1969) The Sharps chondrite – New evidence on the origin of chondrules and chondrites. In *Meteorite Research*. Ed. P. M. Millman. Reidel, pp. 155–65.

Fredriksson K., Nelen J., Melson W. G., Henderson E. P. and Anderson C. A. (1970) Lunar glasses and micro-breccias: Properties and origin. *Science* **167**, 664–6.

Fredriksson K., Noonan A. and Nelen J. (1973) Meteoritic, lunar, and lunar impact chondrules. *Moon* **7**, 475–82.

Fruland R. M., King, E. A. and McKay D. S. (1978) Allende dark inclusions. *Lunar Planet. Sci.* **IX**, 1305–29.

Fujii N. and Miyamoto M. (1983) Constraints on the heating and cooling processes of chondrule formation. In *Chondrules and Their Origins*. Ed. E. A. King. Lunar and Planetary Institute, pp. 53–60.

Fujimaki H., Matsu-ura M., Sunagawa I. and Aoki K. (1981) Chemical compositions of Chondrules and matrices in the ALH-77015 chondrite (L3). *Proceedings of the Sixth Symposium on Antarctic Meteorites*. National Institute of Polar Research, pp. 161–74.

Funaki M., Nagata R. and Momose K. (1981) Natural remanent magnetizations of chondrules, metallic grains and matrix of an Antarctic chondrite, ALH-769. *Mem. Natl. Inst. Polar Res. Special Issue* **20**, 300–15.

Gaffey M. J. and Gilbert S. L. (1998) Asteroid 6 Hebe: The probable parent body of the H-Type ordinary chondrites and the IIE iron meteorites. *Meteorit. Planet. Sci.* **33**, 1281–95.

Gaffey M. J., Burbine T. H. and Binzel R. P. (1993a) Asteroid spectroscopy: Progress and perspective. *Meteoritics* **28**, 161–87.

Gaffey M. J., Bell J. F., Brown R. H., et al. (1993b) Mineralogical variations within the S-type asteroids class. *Icarus* **106**, 573–602.

Georges P., Libourel G. and Deloule E. (2000) Experimental constraints on alkali condensation in chondrule formation. *Meteorit. Planet. Sci.* **35**, 1183–8.

Gibbard S. G. and Levy E. H. (1994) On the possibility of precipitation induced vertical lightning in the protoplanetary nebula (abstract). *Chondrules and the Protoplanetary Disk*, LPI Contrib. 844. Lunar and Planetary Institute, p. 9.

Gilmour J. D., Whitby J. A., Turner G., Bridges J. C. and Hutchison R. (2000) The iodine–xenon system in clasts and chondrules from ordinary chondrites: Implications for early solar system chronology. *Meteorit. Planet. Sci.* **35**, 445–56.

Goldstein J. I. and Short J. M. (1967) The iron meteorites, their thermal history and parent bodies. *Geochim. Cosmochim. Acta* **31**, 1733–70.

Gooding J. L. (1983) Survey of chondrule average properties in H-, L-, and LL-group chondrites: Are chondrules the same in all unequilibrated ordinary chondrites? In *Chondrules and Their Origins*. Ed. E. A. King. Lunar and Planetary Institute, pp. 61–87.

Gooding J. L. and Keil K. (1981) Relative abundances of chondrule primary textural types in ordinary chondrites and their bearing on conditions of chondrule formation. *Meteoritics* **16**, 17–43.

Gooding J. L. and Muenow D. W. (1976) Activated release of alkalis during the vesiculation of molten basalts under high vacuum: Implications for lunar volcanism. *Geochim. Cosmochim. Acta* **40**, 675–86.

  (1977) Experimental vaporization of the Holbrook chondrite. *Meteoritics* **12**, 401–8.

Gooding J. L., Keil K., Fukuoka T. and Schmitt R. A. (1980) Elemental abundances in chondrules from unequilibrated chondrites: Evidence for chondrule origin by melting of preexisting materials. *Earth Planet. Sci. Lett.* **50**, 171–80.

Gooding J. L., Mayeda T. K., Clayton R. N., et al. (1982) Oxygen isotopic compositions of chondrules in unequilibrated chondrites: Further petrological interpretations (abstract). *Lunar Planet. Sci.* **XIII**, 271–2.

Gooding J. L., Mayeda T. K., Clayton R. N. and Fukuoka T. (1983) Oxygen isotopic heterogeneities, their petrological correlations and implicatons for melt origins of chondrules in unequilibrated ordinary chondrites. *Earth Planet. Sci. Lett.* **65**, 209–24.

Göpel C., Manhes G. and Allegre C. J. (1994) U–Pb systematics of phosphates from equilibrated ordinary chondrites. *Earth Planet. Sci. Lett.* **121**, 153–71.

Goswami J. N., Sahijpal S., Kehm K., et al. (1998) In situ determination of iodine content and I–Xe systematics in silicates and troilite phases in chondrules from the LL3 chondrite Semarkona. *Meteorit. Planet. Sci.* **33**, 527–34.

Grabb J. and Schultz L. (1981) Cosmic-ray exposure ages of ordinary chondrites and their significance for parent body stratigraphy, *Geochim. Cosmochim. Acta* **45**, 2151–60.

Graup G. (1981) Terrestrial chondrules, glass spherules and accretionary lapilli from the suevite, Ries Crater, Germany. *Earth Planet. Sci. Lett.* **55**, 407–18.

Gray C. M. and Papanastassiou D. A. (1973) The identification of early condensates from the solar nebula. *Icarus* **20**, 213–39.

Greenberg J. M. (1976) Radical formation, chemical processing, and explosion of interstellar grains. *Astrophys. Space Sci.* **39**, 9–18.

Greenberg R. and Chapman C. R. (1983) Asteroids and meteorites: Parent bodies and delivered samples. *Icarus* **55**, 455–81.

Greenberg R. and Nolan M. C. (1989) Delivery of asteroids and meteorites to the inner solar system. In *Asteroids II*. Ed. R. P Binzel, T. Gehrels and M. S. Matthews. University of Arizona Press, pp. 779–804.

Greenwood J. P. and Hess P C. (1996) Congruent melting kinetics: Constraints on chondrule formation. In *Chondrules and the Protoplanetary Disk*. Ed. R. H. Hewins, R. H. Jones and E. R. D. Scott. Cambridge University Press, pp. 205–11.

Grimm R. E. and McSween H. Y. Jr (1989) Water and thermal evolution of carbonaceous chondrite parent bodies. *Icarus* **82**, 244–80.

   (1993) Heliocentric zoning of the asteroid belt by aluminum-26 heating. *Science* **259**, 653–5.

Grossman J. N. (1985) Chemical evolution of the matrix of Semarkona (abstract). *Lunar Planet. Sci.* **XVI**, 302–3.

   (1988) Formation of chondrules. In *Meteorites and the Early Solar System*. Ed. J. F. Kerridge and M. S. Matthews. University of Arizona Press, pp. 680–96.

   (1996a) The redistribution of sodium in Semarkona chondrules by secondary processes (abstract). *Lunar Planet. Sci.* **XVII**, 467–8.

   (1996b) Chemical fractionations of chondrites: Signatures of events before chondrule formation. In *Chondrules and the Protoplanetary Disk*. Ed. R. H. Hewins, R. H. Jones and E. R. D. Scott. Cambridge University Press, pp. 243–53.

Grossman J. N. and Rubin A. E. (1986) The origin of chondrules and clasts bearing calcic plagioclase in ordinary chondrites (abstract). *Lunar Planet. Sci.* **XVIII**, 293–4.

Grossman J. N. and Wasson J. T. (1982) Evidence for primitive nebular components in chondrules from Chainpur chondrite. *Geochim. Cosmochim. Acta* **46**, 1081–99.

   (1983a) Refractory precursor components of Semarkona chondrules and the fractionation of refractory elements among chondrites. *Geochim. Cosmochim. Acta* **47**, 759–71.

   (1983b) The compositions of chondrules in unequilibrated chondrites: An evaluation of models for the formation of chondrules and their precursor materials. In *Chondrules and Their Origins*. Ed. E. A. King. Lunar and Planetary Institute, pp. 88–121.

   (1985) The origin and history of the metal and sulfide components of chondrules. *Geochim. Cosmochim. Acta* **49**, 925–39.

   (1987) Compositional evidence regarding the origins of rims on Semarkona chondrules. *Geochim. Cosmochim. Acta* **51**, 3003–11.

Grossman J. N., Kracher A. and Wasson J. T. (1979) Volatiles in Chainpur chondrules. *Geophys. Res. Lett.* **6**, 597–600.

Grossman J. N., Rubin A. E., Rambaldi E. R., Rajan R. S. and Wasson J. T. (1985) Chondrules in the Qingzhen type-3 enstatite chondrite: Possible precursor components and comparison to ordinary chondrite chondrules. *Geochim. Cosmochim. Acta* **49**, 1781–95.

Grossman J. N., Rubin A. E., Nagahara H. and King E. A. (1988a) Properties of chondrules. In *Meteorites and the Early Solar System*. Ed. J. F. Kerridge and M. S. Matthews. University of Arizona Press, pp. 619–59.

Grossman J. N., Rubin A. E. and MacPherson G. J. (1988b) ALH85085 – A unique volatile-poor carbonaceous chondrite with possible implications for nebular fractionation processes. *Earth Planet. Sci. Lett.* **91**, 33–54.

Grossman J. N., Alexander C. M. O'D., Wang Jianhua and Brearley A. J. (2000) Bleached chondrules: Evidence for widespread aqueous processes on the parent asteroids of ordinary chondrites. *Meteorit. Planet. Sci.* **35**, 467–86.

Grossman J. N., Alexander C. M. O'D., Wang Jianhua and Brearley A. J. (2002) Zoned chondrules in Semarkona: Evidence for high- and low-temperature processing. *Meteorit. Planet. Sci.* **37**, 49–73.

Grossman L. (1972) Condensation in the primitive solar nebula. *Geochim. Cosmochim. Acta* **36**, 597–619.

Grossman L. and Larimer J. W. (1974) Early chemical history of the solar system. *Rev. Geophys. Space Phys.* **12**, 71–101.

Guimon R. K., Keck B. D., Weeks K. S., DeHart J. and Sears D. W. G. (1985) Chemical and physical studies of type 3 chondrites-IV: Annealing studies of a type 3.4 ordinary chondrite and the metamorphic history of meteorites. *Geochim. Cosmochim. Acta* **49**, 1515–24.

Guimon R. K., Symes S. J. K., Sears D. W. G. and Benoit P. H. (1995) Chemical and physical studies of type 3 chondrites XII: The metamorphic history of CV chondrites and their components. *Meteoritics* **30**, 704–14.

Haack H., Rasmussen K. L. and Warren P. H. (1990) Effects of regolith/megaregolith insulation on the cooling histories of different asteroids, *J. Geophys. Res.* **95**, 5111–24.

Haidinger W. K. (1867) Die Meteoriten des k. k. Hof-Mineraliencabinetes am 1 Juli 1867, und der Fortshritte set 7 Jänuar 1859. *Sitz. Kaiserlichen königlichen. Akad. Wiss. Wien.*

Hamilton P. J., Evensen N. M. and O'Nions R. K. (1979) Chronology and chemistry of Parnallee (LL-3) chondrules (abstract). *Lunar Planet. Sci.* **X**, 494–5.

Harris P. G. and Tozer D. C. (1967) Fractionation of iron in the solar system. *Nature* **215**, 1449–51.

Hartmann L. (1996) Astronomical observations of phenomena in protostellar disks. In *Chondrules and the Protoplanetary Disk*. Ed. R. H. Hewins, R. H. Jones and E. R. D. Scott. Cambridge University Press, pp. 13–20.

Hartmann L. and Kenyon S. J. (1985) On the nature of FU Orionis objects. *Astrophys. J.* **299**, 462–78.

Hartmann W. K., Philips R. J. and Taylor G. J. (1986) *Origin of the Moon*. Lunar and Planetary Institute.

Hashimoto A. (1983) Evaporation metamorphism in the early solar nebula – evaporation experiments on the melt $FeO-MgO-SiO_2-CaO-Al_2O_3$ and chemical fractions of primitive material. *Geochim. J.* **17**, 111–45.

Hashimoto A., Kumazawa M. and Onuma N. (1979) Evaporation metamorphism of primitive dust material in the early solar nebula. *Earth Planet. Sci. Lett.* **43**, 13–21.

Hashimoto A., Davis A. M., Clayton R. N. and Mayeda T. K. (1989) Correlated isotopic mass fractionation of oxygen, magnesium and silicon in forsterite evaporation residues (abstract). *Meteoritics* **24**, 275.

Heide F. and Wlotzka F. (1995) *Meteorites: Messengers from Space*. Springer-Verlag.

Heiken G. H., Vaniman D. T. and French B. M., Eds. (1991) *Lunar Sourcebook: A User's Guide to the Moon*. Cambridge University Press.

Herbig G. H. (1978) Some aspects of early stellar evolution that may be relevent to the origin of the solar system. In *The Origin of the Solar System*. Ed. S. F. Dermott. John Wiley, pp. 219–35.

Herndon J. M. and Herndon M. A. (1977) Aluminum-26 as a planetoid heat source in the early solar system. *Meteoritics* **12**, 459–65.

Herzberg C. T. (1979) The solubility of olivine in basaltic liquid: An ionic model. *Geochim. Cosmochim. Acta* **43**, 1241–51.

Hewins R. H. (1983) Dynamic crystallization experiments as constraints on chondrule genesis. In *Chondrules and Their Origins*. Ed. E. A. King. Lunar and Planetary Institute, pp. 122–33.

(1988) Experimental studies of chondrules. In *Meteorites and the Early Solar System*. Ed. J. F. Kerridge and M. S. Matthews. University of Arizona Press, pp. 660–79.

(1989) The evolution of chondrules. *Proc. NIPR Symp Antarctic Meteorites 2.* National Institute of Polar Research, pp. 200–20.

(1991) Retention of sodium during chondrule formation. *Geochim. Cosmochim. Acta* **55**, 935–42.

(1996) Chondrules and the protoplanetary disk: An overview. In *Chondrules and the Protoplanetary Disk.* Ed. R. H. Hewins, R. H. Jones and E. R. D. Scott. Cambridge University Press, pp. 3–9.

Hewins R. H. and Connolly H. C. Jr (1996) Peak temperatures of flash-melted chondrules. In *Chondrules and the Protoplanetary Disk.* Ed. R. H. Hewins, R. H. Jones and E. R. D. Scott. Cambridge University Press, pp. 197–204.

Hewins R. H. and Newsom H. E. (1988) Igneous activity in the early solar system. In *Meteorites and the Early Solar System.* Ed. J. F. Kerridge and M. S. Matthews. University of Arizona Press, pp. 73–101.

Hewins R. H. and Radomsky P. M. (1990) Temperature conditions of chondrule formation. *Meteoritics* **25**, 309–18.

Hewins R. H., Klein L. C. and Fasano B. V. (1981) Conditions of formation of pyroxene excentroradial chondrules. *Lunar Planet. Sci.* **XII**, 1123–33.

Hewins R. H., Jones R. H. and Scott E. R. D., Eds. (1996) *Chondrules and the Protoplanetary Disk.* Cambridge University Press.

Heymann D. (1967) On the origin of hypersthene chondrites: Ages and shock effects of black chondrites. *Icarus* **6**, 189–221.

Hinton R. W., Long J. V. P., Fallick A. E. and Pillinger C. T. (1983) Ion microprobe measurement of D/H ratios in meteorites (abstract). *Lunar Planet. Sci.* **XIV**, 313–14.

Holder J. and Ryder G. (1995) Unique glass particles from 68001 lunar core thin sections. *Lunar News* **58**, 7.

Hood L. L. and Horányi M. (1991) Gas dynamic heating of chondrule precursor grains in the solar nebula. *Icarus* **93**, 259–69.

(1993) The nebular shock wave model for chondrule formation: one-dimensional calculations. *Icarus* **106**, 179–89.

Hood L. L. and Kring D. A. (1996) Models for multiple heating mechanisms. In *Chondrules and the Protoplanetary Disk.* Ed. R. H. Hewins, R. H. Jones and E. R. D. Scott. Cambridge University Press, pp. 265–76.

Hoppe P., Goswami J. N., Krähenbühl U. and Marti K. (2001) Boron in Chondrules. *Meteorit. Planet. Sci.* **36**, 1331–43.

Horányi M. and Robertson S. (1996) Chondrule formation in lightning discharges: Status of theory and experiments. In *Chondrules and the Protoplanetary Disk.* Ed. R. H. Hewins, R. H. Jones and E. R. D. Scott. Cambridge University Press, pp. 303–10.

Horányi M., Morrill G., Goertz C. K. and Levy E. H. (1985) Chondrule formation in lightning discharges. *Icarus* **114**, 174–85.

Hörz F. and Cintala M. (1997) Impact experiments related to the evolution of planetary regoliths. *Meteorit. Planet. Sci.* **32**, 179–209.

Hörz F. and Schaal R. B. (1981) Asteroid agglutinate formation and implications for asteroid surfaces. *Icarus* **46**, 337–53.

Hörz F., Grieve R., Heiken G., Spudis P. and Binder A. (1991) Lunar surface processes. In *Lunar Sourcebook.* Ed. G. H. Heiken, D. T. Vaniman and B. M. French. Lunar and Planetary Institute, pp. 61–120.

Housen K. R. (1992) Crater ejecta velocities for impacts on rocky bodies (abstract). *Lunar Planet Sci.* **XXIII**, 555–6.

Housen K. R., Wilkening L. L., Chapman C. R. and Greenberg R. (1979) Asteroidal regoliths. *Icarus* **39**, 317–51.

Housley R. M. and Cirlin E. H. (1983) On the alteration of Allende chondrules and the formation of matrix. In *Chondrules and Their Origins*. Ed. E. A. King. Lunar and Planetary Institute, pp. 145–61.

Howard E. C. (1802) Experiments and observations on certain stony substances, which at different times are said to have fallen on the Earth; also on various kinds of native iron. *Phil. Trans.* **92**, 168–212.

Hua X., Adam J., Palme H. and El Goresy A. (1988) Fayalite-rich rims, veins, and halos around and in forsteritic olivines in CAIs and chondrules in carbonaceous chondrites: types, compositional profiles and constraints on their formation. *Geochim. Cosmochim. Acta* **52**, 1389–408.

Huang S., Benoit P. H. and Sears D. W. G. (1993a) Metal and sulfide in Semarkona chondrules and rims: Evidence for reduction, evaporation and recondensation during chondrule formation (abstract). *Meteoritics* **28**, 367–8.

Huang S., Benoit P. H. and Sears D. W. G. (1993b) The group A3 chondrules of Krymka: Further evidence for major evaporative loss during the formation of chondrules (abstract). *Lunar Planet. Sci.* **XXIV**, 681–2.

Huang S., Benoit P. H. and Sears D. W. G. (1994) Group A5 chondrules in ordinary chondrites: their formation and metamorphism (abstract) *Lunar Planet. Sci.* **XXV**, 573–4.

Huang S., Akridge G. and Sears D. W. G. (1996a) Metal–silicate fractionation in the surface dust layers of accreting planetesimals: Implications for the formation of ordinary chondrites and the nature of asteroid surfaces. *J. Geophys. Res. (Planets)* **101**, 29373–85.

Huang S., Lu J., Prinz M., Weisberg M. K., Benoit P. H. and Sears D. W. G. (1996b) Chondrules: Their diversity and the role of open-system processes during their formation. *Icarus* **122**, 316–46.

Hughes D. W. (1978) A disaggregation and thin section analysis of size and mass distributions of the chondrules in the Bjurböle and Chainpur meteorites. *Earth Planet. Sci. Lett.* **38**, 391–400.

Humboldt A. V. (1849) *Cosmos: A sketch of a Physical Description of the Universe 1*. H. G. Bohn, pp. 97–212.

Huss G. R. (1988) The role of presolar dust in the formation of the solar system. *Earth, Moon, Planets* **40**, 165–211.

Huss G. R., Keil K. and Taylor G. J. (1981) The matrices of unequilibrated ordinary chondrites: Implications for the origin and history of chondrites. *Geochim. Cosmochim. Acta* **45**, 33–51.

Huss G. R., MacPherson G. J., Wasserburg G. J., Russell S. S. and Srinivasan G. (2001) ^{26}Al in CAIs and chondrules from unequilibrated ordinary chondrites. *Meteorit. Planet. Sci.* **36**, 975–97.

Hutcheon I. D., Hutchison R. and Wasserburg G. J. (1989) Evidence from the Semarkona ordinary chondrite for ^{26}Al heating of small planets, *Nature* **237**, 238–41.

(1982) Meteorites – Evidence for the interrelationships of materials in the solar system 4.55 Ga ago. *Earth Planet. Sci. Lett.* **29**, 199–208.

Hutchison R. (1983) *The Search for our Beginning*. British Museum (Natural History)/Oxford University Press, p. 39.

(1992) New evidence for the origin of white matrix in Tieschitz (abstract). *Meteoritics* **27**, 236–7.

(1996). Chondrules and their associates in ordinary chondrites: A planetary connection? In *Chondrules and the Protoplanetary Disk*. Ed. R. H. Hewins, R. H. Jones and E. R. D. Scott. Cambridge University Press, pp. 311–18.

Hutchison R. and Bevan A. W. R. (1983) Conditions and time of chondrule accretion. In *Chondrules and Their Origins*. Ed. E. A. King. Lunar and Planetary Institute, pp. 162–79.

Hutchison R., Bevan A. W. R., Agrell S. O. and Ashworth J. R. (1979) Accretion temperature of the Tieschitz, H3, chondritic meteorite. *Nature* **280**, 116–19.

Hutchison R., Alexander C. M. O. and Barber D. J. (1987) The Semarkona meteorite: First recorded occurrence of semecite in an ordinary chondrite, and its implications. *Geochim. Cosmochim. Acta* **51**, 1875–82.

Hutchison R., Alexander C. M. O. and Barber D. J. (1988). Chondrules: Chemical, mineralogical and isotopic constraints on theories of their origin. *Phil. Trans. Roy. Soc. (London)* **A325**, 445–58.

Ihinger P. D. and Stolper E. (1986) The color of meteoritic hibonite: An indicator of oxygen fungacity. *Earth Planet. Sci. Lett.* **78**, 67–79.

Ikeda Y. (1983) Major element compositions and chemical types of chondrules in unequilibrated E, O, and C chondrites from Antarctica. *Mem. Natl. Inst. Polar Res., Spec. Issue* **30**, 122–45.

  (1989) Petrochemical study of the Yamato-691 enstatite chondrite (E3) V: Comparison of major element chemistries of chondrules and inclusions in Y-691 with those in ordinary and carbonaceous chondrites. *Proceedings of the NIPR Symp. Antarctic Meteorites 2*. National Institute of Polar Research, pp. 147–65.

Ikeda Y. and Kimura M. (1985) Na–Ca zoning of chondrules in Allende and ALHA-77003 carbonaceous chondrites. *Meteoritics* **20**, 670–1.

Irving A. J. (1978) A review of experimental studies of crystal/liquid trace element partitioning. *Geochim. Cosmochim. Acta* **42**, 743–70.

Ishii T., Miyamoto M. and Takeda H. (1976) Pyroxene geothermometry and crystallization-, subsolidus equilibration-temperatures of lunar and achondritic pyroxenes. *Lunar Sci.* **VII**, 408–10.

Ivanov A. V., Zolensky M. E., Brandstätter F., Kurat G. and Kononkova N. N. (1994) A phyllosilicate–sulfide vein in Kaidun (abstract). *Meteoritics* **29**, 477.

Jarosewich E. (1990) Chemical analyses of meteorites: A compilation of stony and iron meteorite analyses. *Meteoritics* **25**, 323–37.

Jarosewich E. and Dodd R. T. (1981) Chemical variations among L-chondrites – II: Chemical distinctions between L3 and LL3 chondrites. *Meteoritics* **16**, 83–91.

  (1985) Chemical variations among L-chondrites – IV: Analyses, with petrographic notes, of 13 L-group and 3-LL group chondrites. *Meteoritics* **20**, 23–36.

Johannes, W. (1968) Experimental investigation of the reaction forsterite + H_2O = serpentine + brucite, *Contrib. Mineral. Petrol.* **19**, 309–15.

Johnson C. A., Prinz M., Weisberg M. K., Clayton R. N. and Mayeda T. K. (1990) Dark inclusions in Allende, Leoville, and Vigarano – Evidence for nebular oxidation of CV3 constituents. *Geochim. Cosmochim. Acta* **54**, 819–30.

Johnson M. C. (1986) The solar nebula redox state as recorded by the most reduced chondrules of five primitive chondrites. *Geochim. Cosmochim. Acta* **50**, 1497–502.

Jones R. H. (1990) Petrology and mineralogy of type II chondrules in Semarkona (LL3.0): Origin of closed-system fractional crystallization, with evidence for supercooling. *Geochim. Cosmochim. Acta* **54**, 1785–802.

  (1992) On the relationship between isolated and chondrule olivine grains in the carbonaceous chondrite ALHA77307. *Geochim. Cosmochim. Acta* **56**, 467–82.

  (1994) Petrology of FeO-poor, porphyritic pyroxene chondrules in the Semarkona chondrite. *Geochim. Cosmochim. Acta* **58**, 5325–40.

(1996) Relict grains in chondrules: Evidence for chondrule recycling. In *Chondrules and the Protoplanetary Disk*. Ed. R. H. Hewins, R. H. Jones and E. R. D. Scott. Cambridge University Press, pp. 163–72.

Jones R. H. and Danielson L. R. (1997) A chondrule origin for dusty relict olivine in unequilibrated chondrites. *Meteorit. Planet. Sci.* **32**, 753–60.

Jones R. H. and Lofgren G. E. (1993) A comparison of FeO-rich, porphyritic olivine chondrules in unequilibrated chondrites and experimental analogues. *Meteoritics* **28**, 213–21.

Jones R. H. and Scott E. R. D. (1989) Petrology and thermal history of type IA chondrules in the Semarkona (LL3.0) chondrite. *Proc. 19th Lunar Planet. Sci. Conf.* Lunar and Planetary Institute, pp. 523–36.

Jurewicz A. J. G. and Watson E. B. (1988) Cations in olivine part I: Calcium partitioning and calcium–magnesium distribution between olivines and coexisting melts, with petrologic applications. *Contrib. Mineral. Petrol.* **99**, 176–85.

Kallemeyn, G. W. (1988) Elemental variations in bulk chondrites: A brief review. In *Meteorites and the Early Solar System*. Ed. J. F. Kerridge and M. S. Matthews. University of Arizona Press, pp. 390–3.

Kallemeyn G. W. and Wasson J. T. (1982a). The compositional classification of chondrites: III. Ungrouped carbonaceous chondrites. *Geochim. Cosmochim. Acta.* **49**, 2217–28.

(1982b). The compositional classification of chondrites: IV. Ungrouped chondritic meteorites and clasts. *Geochim. Cosmochim. Acta.* **49**, 261–70.

Kallemeyn G. W., Rubin A. E. and Wasson J. T. (1991) The compositional classification of chondrites: V. The Karoonda (CK) group carbonaceous chondrites. *Geochim. Cosmochim. Acta* **55**, 881–92.

(1994) The compositional classification of chondrites: VI. The CR carbonaceous chondrite group. *Geochim. Cosmochim. Acta.* **58**, 2873–88.

(1996) The compositional classification of chondrites: VII. The R chondrite group. *Geochim. Cosmochim. Acta.* **60**, 2243–56.

Kaula W. M. (1979) Thermal evolution of Earth and Moon growing by planetesimal impacts. *J. Geophys. Res.* **84**, 999–1008.

Kaushal S. K. and Wetherill G. W. (1969) Rb^{87}–Sr^{87} age of bronzite (H group) chondrites. *J. Geophys. Res.* **74**, 2717–26.

Keil K. (1968) Mineralogical and chemical relationships among enstatite chondrites. *J. Geophys. Res.* **73**, 6945–76.

Keil K. (1982) Composition and origin of chondritic breccias. In *Workshop on Lunar Breccias and Soil and Their Meteoritic Analogs*. Ed. G. J. Taylor and L. L. Wilkening. LPI Tech. Report 82–02. Lunar and Planetary Institute, pp. 65–83.

Keil K. (1989) Enstatite meteorites and their parent bodies. *Meteoritics* **24**, 195–208.

Keil K. and Fredriksson K. (1964) The iron, magnesium, and calcium distribution in coexisting olivines and rhombic pyroxenes of chondrites. *J. Geophys. Res.* **69**, 3487–515.

Keil K., Kurat G., Prinz M. and Green J. A. (1972) Lithic fragments, glasses and chondrules from Luna 16 finds. *Earth Planet. Sci. Lett.* **13**, 243–56.

Keil K., Prinz M., Planner H. N., *et al.* (1973) A qualitative comparison of textures in lunar chondrules and CO_2 laser-formed synthetic chondrule-like spherules. *Institution of Meteoritics Special Publication No. 7.* University of New Mexico.

Keller L. P. and Buseck P. R. (1990) Matrix mineralogy of Lance CO3 carbonaceous chondrite: A transmission electron microscope study. *Geochim. Cosmochim. Acta* **54**, 1155–63.

Kelly W. R. and Larimer J. W. (1977) Chemical fractionations in meteorites. VIII. Iron meteorites and the cosmochemical history of the metal phase. *Geochim. Cosmochim. Acta* **41**, 93–111.
Kerridge J. F. (1964) Low-temperature minerals from the fine-grained matrix of some carbonaceous chondrites. *Ann. N. Y. Acad. Sci.* **119**, 41–53.
  (1993) What can meteorites tell us about nebular conditions and processes during planetesimal accretion? *Icarus* **106**, 135–50.
Kerridge J. F. and Bunch T. E. (1979) Aqueous activity on asteroids: Evidence from carbonaceous chondrites. In *Asteroids*. Ed. T. Gehrels. University of Arizona Press, pp. 745–64.
Kerridge J. F. and Kieffer S. W. (1977). A constraint on impact theories of chondrule formation. *Earth Planet. Sci. Lett.* **35**, 35–42.
Kerridge J. F. and Matthews M. S. (1988) *Meteorites and the Early Solar System*. University of Arizona Press.
Kieffer S. W. (1975). Droplet chondrules. *Science* **189**, 333–40.
Kimura M. and Watanabe S. (1986). Adhesive growth and abrasion of chondrules during the accretion process. *Mem. Natl. Inst. Polar Res., Spec. Issue* **41**, 222–34.
Kimura M. and Yagi K. (1980) Crystallization of chondrules in ordinary chondrites. *Geochim. Cosmochim. Acta* **44**, 589–602.
King E. A. (1982). Refractory residues, condensates and chondrules from solar furnace experiments. *Proc. 13th Lunar Planet. Sci. Conf. J. Geophys. Res.* **87**, A429–34.
  (Ed.) (1983a) *Chondrules and Their Origins*. Lunar and Planetary Institute.
  (1983b) Reduction, partial evaporation, and spattering: Possible chemical and physical processes in fluid drop chondrule formation. In *Chondrules and Their Origins*. Ed. E. A. King. Lunar and Planetary Institute, pp. 180–7.
King E. A., Carman M. F. and Butler J. C. (1972a) Chondrules in Apollo 14 samples: Implications for the origin of chondritic meteorites. *Science* **195**, 59–60.
King E. A., Carman M. F. and Butler J. C. (1972b) Chondrules in Apollo 14 samples and size analyses of Apollo 14 and 15 finds. *Proc. III Lunar Sci. Conf.* Lunar and Planetary Institute, pp. 673–86.
King T. V. V. and King E. A. (1978) Grain size and petrography of C2 and C3 carbonaceous chondrites. *Meteoritics* **13**, 47–72.
  (1979) Size–frequency distributions of fluid drop chondrules in ordinary chondrites. *Meteoritics* **14**, 91–6.
  (1981) Accretionary dark rims in unequilibrated ordinary chondrites. *Icarus* **48**, 460–72.
Kitamura M. and Tsuchiyama A. (1996) Collision of icy and slightly differentiated bodies as an origin for unequilibrated ordinary chondrites. In *Chondrules and the Protoplanetary Disk*. Ed. R. H. Hewins, R. H. Jones and E. R. D. Scott. Cambridge University Press, pp. 319–26.
Klein C. (1906) *Studien uber Meteoriten*, p. 35. (Cited in Merrill, 1920.)
Koeberl C., Kurat G. and Brandstätter F. (1991) MAC 88105 – A regolith breccia from the lunar highlands: Mineralogical, petrological, and geochemical studies. *Geochim. Cosmochim. Acta* **55**, 3073–87.
Kozul J. M., Ulmer G. C. and Hewins R. H. (1988) Intrinsic oxygen fugacity measurements of some Allende type B inclusions. *Geochim. Cosmochim. Acta* **52**, 2107–16.
Kracher A., Scott E. R. D. and Keil K. (1984) Relict and other anomalous grains in chondrules; Implications for chondrule formation. *Proc. XIV Lunar Planet. Sci. Conf. J. Geophys. Res.* **89**, B559–66.

Kring D. A. (1986) O/H in the solar nebula gas in the zones of the C2, C3, and UOC chondrule formation (abstract). *Lunar Planet. Sci.* **XVII**, 450–1.

Kring D. A. (1987) Fe, Ca-rich rims around magnesian chondrules in the Kainsaz (CO3) chondrite (abstract). *Lunar Planet. Sci.* **XVIII**, 517–18.

Kring D. A. and Wood J. A. (1987) Fe, Ca-rich and Mg-rich chondrule rims in the Kainsaz (CO3) chondrite: Evidence of fluctuating nebular conditions (abstract). *Meteoritics* **22**, 432.

Krot A. N. and Keil K. (2002) Anorthite-rich chondrules in CR and CH carbonaceous chondrites: Genetic link between Ca, Al-rich inclusions and ferromagnesian chondrules. *Meteorit. Planet. Sci.* **37**, 91–111.

Krot A. N. and Rubin A. E. (1996) Microchondrule-bearing chondrule rims: Constraints on chondrule formation. In *Chondrules and the Protoplanetary Disk*. Ed. R. H. Hewins, R. H. Jones and E. R. D. Scott. Cambridge University Press, pp. 173–84.

Krot A. N., Petaev M. I., Scott E. R. D., *et al.* (1998) Progressive alteration in CV3 chondrites: More evidence for asteroid alteration. *Meteorit. Planet. Sci.* **33**, 1033–40.

Krot A. N., Ulyanov A. A., Meibom A. and Keil K. (2001) Forsterite-rich accretionary rims around Ca, Al-rich inclusions from the reduced CV3 chondrite Efremovka. *Meteorit. Planet. Sci.* **36**, 611–28.

Krot A. N., Hutcheon I. D. and Keil K. (2002) Plagioclase-rich chondrules in the reduced CV chondrites: Evidence for complex formation history and genetic links between calcium–aluminum-rich inclusions and ferromagnesian chondrules. *Meteorit. Planet. Sci.* **37**, 155–82.

Kunii D. and Levenspiel O. (1991) *Fluidization Engineering*, 2nd edn. Butterworth and Heinemann.

Kurat G., Keil K., Prinz M. and Nehru C. E. (1972) Chondrules of lunar origin. *Proc. 3rd Lunar Sci. Conf.* part 1. Lunar and Planetary Institute, pp. 707–21.

Kurat G., Keil K. and Prinz M. (1974) Rock 14318: a polymict lunar breccia with chondritic texture. *Geochim. Cosmochim. Acta* **38**, 1133–46.

Lange D. E. and Larimer J. W. (1973) Chondrules: an origin by impacts between dust grains. *Science* **182**, 920–2.

Lange M. A. and Ahrens T. J. (1982) The evolution of an impact-generated atmosphere. *Icarus* **51**, 96–120.

Langevin Y. and Maurette M. (1980) A model for small body regolith evolution: the critical parameters (abstract). *Lunar Planet. Sci.* **XI**, 602–4.

Lanoix M., Strangway D. W. and Pearce G. W. (1977) Anomalous acquisition of thermoremanence at 130 °C in iron and paleointensity of the Allende meteorite. *Lunar Planet. Sci.* **VIII**, 689–701.

  (1978) The primordial magnetic field preserved in chondrules of the Allende meteorite. *Geophys. Res. Lett.* **5**, 73–6.

Larimer J. W. (1967) Chemical fractionations in meteorites, I. Condensation of the elements. *Geochim. Cosmochim. Acta* **37**, 1603–23.

  (1988) The cosmochemical classification of the elements. In *Meteorites and the Early Solar System*. Ed. J. F. Kerridge and M. S. Matthews. University of Arizona Press, pp. 375–89.

Larimer J. W. and Anders E. (1967) Chemical fractionation in meteorites-II. Abundance patterns and their intrepretation. *Geochim. Cosmochim. Acta* **31**, 1239–70.

  (1970) Chemical fractionation in meteorites-III. Major element fractions in chondrites. *Geochim. Cosmochim. Acta* **34**, 367–87.

Larimer J. W. and Bartholomay M. (1979) The role of carbon and oxygen in cosmic gases – Some applications to the chemistry and mineralogy of enstatite chondrites. *Geochim. Cosmochim. Acta* **43**, 1455–66.

Larimer J. W. and Wasson J. T. (1988a) Refractory lithophile elements. In *Meteorites and the Early Solar System*. Ed. J. F. Kerridge and M. S. Matthews. University of Arizona Press, pp. 394–415.

(1988b) Siderophile element fractionation. In *Meteorites and the Early Solar System*. Ed. J. F. Kerridge and M. S. Matthews. University of Arizona Press, pp. 416–35.

Larson H. P. and Veeder G. J. (1979) Infrared spectral reflectances of asteroid surfaces. In *Asteroids*. Ed. T. Gehrels. University of Arizona Press, pp. 724–44.

Laul J. C., Ganapathy R., Anders E. and Morgan J. W. (1973) Chemical fractionations in meteorites – VI. Accretion temperatures of H-, LL- and E-chondrites from abundance of volatile trace elements. *Geochim. Cosmochim. Acta* **36**, 329–57.

Lavoisier A. (1772) Sur un effect singular de tonnerre. Observations sur la physics, sur la histoire naturelle, et sur les Arts. *J. Physique* **2**, 310–12 (printed and dated 1777).

Lebofsky L. A., Jones T. D. and Herbert E. (1989) Asteroid volatile inventories. In *Origin and Evolution of Planetary and Satellite Atmospheres*. Ed. S. K. Atreya, J. B. Pollack and M. S. Matthews. University of Arizona Press, pp. 192–229.

Lee T. (1988) Implications of isotopic anomalies for nucleosynthesis. In *Meteorites and the Early Solar System*. Ed. J. F. Kerridge and M. S. Matthews. University of Arizona Press, pp. 1063–89.

Lee T., Papanastassiou, D. A. and Wasserburg, G. J. (1976) Demonstration of ^{26}Mg excession in Allende and evidence for ^{26}Al. *Geophys. Res. Lett.* **3**, 41–4.

Lee T., Mayeda T. and Clayton R. N. (1980) Oxygen isotopic anomalies in Allende inclusion HAL. *Geophys. Res. Lett.* **7**, 493–6.

Leitch C. A. and Smith J. V. (1982) Petrography, mineral chemistry and origin of type I enstatite chondrites. *Geochim. Cosmochim. Acta.* **46**, 2083–96.

Levy E. H. (1988) Energetics of chondrule formation. In *Meteorites and the Early Solar System*. Ed. J. F. Kerridge and M. S. Matthews. University of Arizona Press, pp. 697–711.

Levy E. H. and Araki S. (1989) Magnetic reconnection flares in the protoplanetary nebula and the possible origin of meteorite chondrules. *Icarus* **81**, 74–91.

Levy E. H. and Sonett C. P. (1978) Meteorite magnetism and early solar system magnetic fields. In *Protostars and Planets* Ed. T. Gehrels. University of Arizona Press, pp. 516–32.

Lewis J. S. (1976a) Low-temperature condensation from the solar nebula. *Icarus* **16**, 241–52.

(1976b) Metal/silicate fractionation in the solar system. *Earth Planet. Sci. Lett.* **15**, 286–90.

Lewis R. D., Lofgren G. E., Franzen H. F. and Windom K. E. (1993) The effect of Na vapor on the Na content of chondrules. *Meteoritics* **28**, 622–8.

Li Chunlai, Bridges J. C., Hutchison R., et al. (2000). Bo Xian (LL3.9): Oxygen-isotopic and mineralogical characterisation of separated chondrules. *Meteorit. Planet. Sci.* **35**, 561–8.

Liffman K. (1992) The formation of chondrules via ablation. *Icarus* **100**, 608–20.

Liffman K. and Brown M. J. I. (1996) The protostellar jet model of chondrule formation. In *Chondrules and the Protoplanetary Disk*. Ed. R. H. Hewins, R. H. Jones and E. R. D. Scott. Cambridge University Press, pp. 285–302.

Lipschutz M. E. and Woolum D. S. (1988) Highly labile elements. In *Meteorites and the Early Solar System*. Ed. J. F. Kerridge and M. S. Matthews. University of Arizona Press, pp. 462–87.

Lipschutz M. E., Gaffey M. J. and Pellas P. (1989) Meteoritic parent bodies – Nature, number, size and relation to present-day asteroids. In *Asteroids II*. Ed. R. P Binzel, T. Gehrels and M. S. Matthews. University of Arizona Press, pp. 740–77.

Lofgren G. E. (1989) Dynamic crystallization of chondrule melts of porphyritic olivine composition; textures experimental and natural. *Geochim. Cosmochim. Acta* **53**, 461–70.

(1996) A dynamic crystallization model for chondrule melts. In *Chondrules and the Protoplanetary Disk*. Ed. R. H. Hewins, R. H. Jones and E. R. D. Scott. Cambridge University Press, pp. 187–96.

Lofgren G. and Russell W. J. (1986) Dynamic crystallization of chondrule melts of porphyritic and radial pyroxene composition. *Geochim. Cosmochim. Acta* **50**, 1715–26.

Lord H. C. (1965) Molecular equilibrium and condensation in a solar nebula and cool stellar atmospheres. *Icarus* **4**, 279–88.

Love S. G., Keil K. and Scott E. R. D. (1994) Formation of chondrules by electrical discharge heating. In *Papers Presented to Chondrules and the Protoplanetary Disk*, LPI Contrib. 844. Lunar and Planetary Institute, p. 21.

Loveland W., Schmitt R. A. and Fisher D. E. (1969) Aluminum abundances in stony meteorites. *Geochim. Cosmochim. Acta* **33**, 375–85.

Lovering J. F. F., Nichiporuk W., Chodos A. and Brown H. (1957) The distribution of gallium, germaniuun, cobalt, chromium, and copper in iron and stony-iron meteorites in relation to nickel content and structure. *Geochim. Cosmochim Acta* **11**, 263–78.

Lu J. (1992) The physical and chemical studies of chondrules from the type 3 ordinary chondrites. Ph.D. Thesis, University of Arkansas, Fayetteville, AR.

Lu J., Sears D. W. G., Keck B., Prinz M., Grossman J. N. and Clayton R. N. (1990) Semarkona type I chondrules compared with similar chondrules in other classes (abstract). *Lunar Planet. Sci.* **XXI**, 720–1.

Lu J., Sears D. W. G., Benoit P. H., Prinz M. and Weisberg M. K. (1992) The four primitive chondrule groups and the formation of chondrules (abstract). *Lunar Planet. Sci.* **XXIII**, 813–14.

Lux G., Keil K. and Taylor G. J. (1980) Metamorphism of the H-group chondrites: Implications from compositional and textural trends in chondrules. *Geochim. Cosmochim. Acta* **44**, 841–55.

(1981) Chondrules in H3 chondrites: Textures, compositions and origins. *Geochim. Cosmochim. Acta* **45**, 675–85.

Macdougall J. D. and Kothari B. K. (1976) Formation chronology for C2 meteorites. *Earth Planet. Sci. Lett.* **33**, 33–44.

MacPherson G. J., Hashimoto A. and Grossman L. (1985) Accretionary rims on inclusions in the Allende meteorite. *Geochim. Cosmochim. Acta* **49**, 2267–79.

MacPherson G. J., Wark D. A. and Armstrong J. T. (1988) Primitive material surviving in chondrites: Refractory inclusions. In *Meteorites and the Early Solar System*. Ed. J. F. Kerridge and M. S. Matthews. University of Arizona Press, pp. 746–807.

MacPherson G. J., Davis A. M. and Zinner E. K. (1995) The distribution of aluminum-26 in the early Solar System – a reappraisal. *Meteoritics* **30**, 365–86.

Marti K. and Graf T. (1992) Cosmic-ray exposure history of ordinary chondrites. *Ann. Rev. Earth Planet. Sci.* **30**, 244–68.

Martin P. M. and Mills A. A. (1976) Size and shape of chondrules in the Bjurbole and Chainpur meteorites. *Earth and Planet. Sci. Lett.* **33**, 239–48.

(1978) Size and shape of near-spherical Allegan chondrules. *Earth Planet. Sci. Lett.* **38**, 385–90.

(1980) Preferred chondrule orientations in meteorites. *Earth Planet. Sci. Lett.* **51**, 18–25.

Martin P. M., Mills A. A. and Walker E. (1975) Preferential orientation in four C3 chondritic meteorites. *Nature* **257**, 37–8.

Marvin U. B. (1996) Ernst florens Friedrich Chladni (1756–1827) and the origins of modern meteorite research. *Meteoritics* **31**, 545–88.

Marvin U. B., Wood J. A. and Dickey J. S. (1970) Ca-Al rich phases in the Allende meteorite. *Earth Planet. Sci. Lett.* **7**, 346–50.

Mason B. (1960) Origin of chondrules and chondritic meteorites. *Nature* **186**, 230–1.

Mason B. (1962) *Meteorites*. John Wiley.

Mason B. and Taylor S. R. (1982) Inclusions in the Allende meteorite. *Smithsonian Contrib. Earth Sci.* **25**, 1–30.

Masursky H., Batson R. M., Melauley J. F., *et al.* (1972) Mariner 9 television reconnaissance of Mars and its satellites. *Science* **175**, 294–305.

Matsui T. and Osaka M. (1979) Thermal property measurement of Yamato meteorites. *Mem. Nat. Inst. Polar Res. Spec. Issue* **15**, 243–52.

Matsunami S. (1984) The chemical compositions and textures of matrices and chondrule rims of eight unequilibrated ordinary chondrites; A preliminary report. *Mem. Nat. Inst. Polar Res. Spec. Issue* **35**, 126–45.

Matsunami S., Ninagawa K., Nishimura S., *et al.* (1993) Thermoluminescence and compositional zoning in the mesostasis of a Semarkona group A1 chondrule and new insights into the chondrule-forming process. *Geochim. Cosmochim. Acta* **57**, 2102–10.

Matza S. D. and Lipschutz M. E. (1977) Volatile/mobile trace elements in Karoonda (C4) chondrite. *Geochim. Cosmochim. Acta* **41**, 1398–401.

Maurette M. (1993) *Hunting for Stars*. McGraw-Hill.

Mayeda T. K., Clayton R. N. and Sodonis A. (1989) Internal oxygen isotope variations in two unequilibrated chondrites (abstract). *Meteoritics* **24**, 301.

McCall G. J. H. (1973) *Meteorites and Their Origins*. David and Charles.

McCord T. B., Adams J. B. and Johnson T. V. (1970) Asteroid Vesta: Spectral reflectivity and compositional implications. *Science* **168**, 1445–7.

McCoy T. J., Scott E. R. D., Jones R. H., Keil K. and Taylor G. J. (1991) Composition of chondrule silicates in LL3–5 chondrites and implications for their nebular history and parent body metamorphism, *Geochim. Cosmochim. Acta* **55**, 601–19.

McCoy T. J., Keil K., Mayeda T. K. and Clayton R. N. (1992) Monument Draw and the formation of the Acapulcoites. *Lunar Planet. Sci.* **XXIII**, 871–2.

McCoy T. J., Keil K., Ash R. D., *et al.* (1993). Roosevelt County 075: A petrologic chemical and isotopic study of the most unequilibrated known H chondrite. *Meteoritics* **28**, 681–91.

McKay D. S., Swindle T. D. and Greenberg R. (1989) Asteroidal regoliths – What we do not know. In *Asteroids II*. Ed. R. P Binzel, T. Gehrels and M. S. Matthews. University of Arizona Press, pp. 921–45.

McKay D. S., Heiken G., Basu A., *et al.* (1991) The lunar regolith. In *Lunar Sourcebook: A User's Guide to the Moon*. Ed. G. H. Heiken, D. T. Vaniman and B. M. French. Cambridge University Press, pp. 285–356.

McMahon B. M. and Haggerty S. E. (1980) Experimental studies bearing on the magnetite alloy-sulfide association in the Allende meteorite: Constraints on the conditions of chondrule formation. *Proc. 11th Lunar Planet. Sci. Conf.* Lunar and Planetary Institute, pp. 1003–25.

McSween H. Y. Jr (1977a) Carbonaceous chondrites of the Ornans type: a metamorphic sequence. *Geochim. Cosmochim. Acta* **41**, 477–91.

(1977b) Chemical and petrographic constraints on the origin of chondrules and inclusions in carbonaceous chondrites. *Geochim. Cosmochim. Acta* **41**, 1843–60.

(1977c) Petrographic variations among carbonaceous chondrites of the Vigarano type. *Geochim. Cosmochim. Acta* **41**, 477–91.

(1977d) *Chemical analyses of chondrules and inclusions in chondrite meteorites.* Harvard/Smithsonian Center for Astrophysics Report, Cambridge, MA.

(1978) *Chemical analyses of chondrules and inclusions in chondritic meteorites.* Unpublished document available from the author.

(1979a) Are carbonaceous chondrites primitive or processed? A review. *J. Geophys. Space Phys.* **17**, 1059–78.

(1979b) Alteration in CM carbonaceous chondrites inferred from modal and chemical variations in matrix. *Geochim. Cosmochim. Acta* **43**, 1761–70.

(1985) Constraints on chondrule origin from petrology of isotopically characterized chondrules in the Allende meteorite. *Meteoritics* **20**, 523–40.

(1987) *Meteorites and Their Parent Planets.* Cambridge University Press.

McSween H. Y. Jr and Richardson S. M. (1977) The compositions of carbonaceous chondrite matrix. *Geochim. Cosmochim. Acta* **41**, 1145–61.

McSween H. Y. Jr, Fronabarger A. K. and Driese S. G. (1983) Ferromagnesian chondrules in carbonaceous chondrites. In *Chondrules and Their Origins* Ed. E. A. King. Lunar and Planetary Institute, pp. 195–210.

McSween H. Y. Jr, Sears D. W. G. and Dodd R. T. (1988) Thermal metamorphism. In *Meteorites and the Early Solar System.* Ed. J. F. Kerridge and M. S. Matthews. University of Arizona Press, pp. 102–13.

Meibom A. and Clark B. E. (1999) Evidence for the insignificance of ordinary chondritic material in the asteroid belt. *Meteorit. Planet. Sci.* **34**, 7–24.

Melosh H. J. (1989) *Impact Cratering, A Geologic Process.* Oxford University Press.

Merrill G. P. (1920) On chondrules and chondritic structure in meteorites. *Proc. Natl. Acad. Sci.* **6**, 449–72.

(1921) On metamorphism in meteorites. *Geol. Soc. Araer. Bull.* **32**, 395–414.

Metzler K. and Bischoff A. (1996) Constraints on chondrite agglomeration from fine-grained chondrule rims. In *Chondrules and the Protoplanetary Disk.* Ed. R. H. Hewins, R. H. Jones and E. R. D. Scott. Cambridge University Press, pp. 153–61.

Metzler K., Bischoff A. and Stöffler D. (1992) Accretionary dust mantles on CM chondrites: Evidence for solar nebula processes. *Geochim. Cosmochim. Acta* **56**, 2873–97.

Meunier S. (1883) *C. R. Paris Acad. Sci.* **96**. (Cited in Merrill, 1920.)

Minster P. M. and Allégre C. J. (1979) ^{87}Rb–^{87}Sr dating of L chondrites: Effects of shock and brecciation. *Meteoritics* **14**, 235–48.

Misawa K. and Fujita T. (2000) Magnesium isotopic fractionations in barred olivine chondrules from the Allende meteorite. *Meteorit. Planet. Sci.* **35**, 85–94.

Misawa K. and Nakamura N. (1988) Highly fractionated rare-earth elements in ferromagnesian chondrules from the Felix (CO3) meteorite. *Nature* **334**, 47–50.

(1996) Origin of refractory precursor components of chondrules from carbonaceous chondrites. In *Chondrules and the Protoplanetary Disk.* Ed. R. H. Hewins, R. H. Jones and E. R. D. Scott. Cambridge University Press, pp. 99–105.

Miyamoto M., Fujii N. and Takeda H. (1981) Ordinary chondrite parent body: An internal heating model, *Proc. 12th Lunar Planet. Sci. Conf.* Lunar and Planetary Institute, pp. 1145–52.

Miyamoto M., McKay D. S., McKay G. A. and Duke M. B. (1986) Chemical zoning and homogenization of olivines in ordinary chondrites and implications for thermal histories of chondrules. *J. Geophys. Res.* **91**, 12804–16.

Morfill G. (1983) Some cosmochemical consequences of a turbulent proto-planetary cloud. *Icarus* **53**, 41–54.

Morfill G., Spruit, H. and Levy E. H. (1993) Physical processes and conditions associated with the formation of protoplanetary disks. In *Protostars and Planets III*. Ed. E. H. Levy and J. I. Lunine. University of Arizona Press, pp. 939–78.

Morfill G. E., Durisen R. H. and Turner G. W. (1998) An accretion rim constraint on chondrule formation theories. *Icarus* **134**, 180–4.

Morse A. D., Sears D. W. G., Hutchison R., et al. (1988) Alteration of type 3 ordinary chondrites (abstract). *Meteoritics* **23**, 291.

Mostefaoui S., Lugmair G. W., Hoppe P. and El Goresy A. (2002) Evidence for Live Iron-60 in Semarkona and Chervony Kut: A NanoSIMS Study (abstract). *Lunar Planet. Sci.* Abstract no. 1585.

Müller O., Baedecker P. A. and Wasson J. T. (1971) Relationship between siderophile element content and oxidation state of ordinary chondrites. *Geochim. Cosmochim. Acta* **35**, 1121–37.

Murchie S. and Erard S. (1996) Spectral properties and heterogeneity of Phobos from measurements of Phobos 2. *Icarus* **123**, 63–86.

Myson B. O. and Kushiro I. (1988) Condensation, evaporation, melting, and crystallization in the primitive solar nebula; Experimental data in the system $MgO-SiO_2-H_2$ to 1.0×10^{-9} bar and $2870\,°C$ with variable oxygen fugacity. *Amer. Mineral.* **73**, 1–19.

Nagahara H. (1981) Evidence for secondary origin of chondrules. *Nature* **292**, 135–6.

(1983a) Texture of chondrules. *Mem. Natl. Inst. Polar Res. Special Issue* **30**, 61–83.

(1983b) Chondrules formed through incomplete melting of the pre-existing mineral clusters and the origin of chondrules. In *Chondrules and Their Origins*. Ed. E. A. King. Lunar and Planetary Institute, pp. 211–22.

(1984) Matrices of type 3 ordinary chondrites; Primitive nebular records. *Geochim. Cosmochim. Acta* **48**, 2581–95.

(1986) Reduction kinetics of olivine and oxygen fugacity environment during chondrule formation. *Lunar Planet. Sci.* **XVII**, 595–6.

Nagahara H. and Kushiro I. (1987) Origin of iron-rich olivine in the matrices of type 3 ordinary chondrites – an experimental study. *Earth Planet. Sci. Lett.* **85**, 537–47.

(1989) Vaporization experiments in the system plagioclase–hydrogen. *Proceedings of the NIPR Symposium on Antarctic Meteorites*, volume 2, pp. 235–51.

Nagahara H., Kushiro I., Mysen B. O. and Mori H. (1989a) Experimental vaporization and condensation of olivine solid solution. *Nature* **331**, 516–18.

Nagahara H., Kushiro I. and Tomeoka K. (1989b) Vaporization experiments in the system plagioclase–hydrogen: 2. Composition of the gas and residue (abstract). *14th Symposium on Antarctic Meteorites*. National Institute of Polar Research, p. 84.

Nagahara H., Mysen B. O. and Kushiro I. (1994) Evaporation of olivine – low pressure phase relations of the olivine system and its implication for the origin of chondritic components in the solar nebula. *Geochim. Cosmochim. Acta* **58**, 1951–63.

Nagata T. and Funaki M. (1983) Paleointensity of the Allende carbonaceous chondrite. *Mem. Natl. Inst. Polar Res. Special Issue* **30**, 403–34.

Nakamura N. (1974) Determination of REE, Ba, Fe, Mg, Na and K in carbonaceous and ordinary chondrites. *Geochim. Cosmochim. Acta* **38**, 757–75.

Nakamura N. and Masuda A. (1973) Chondrites with peculiar rare-earth patterns. *Earth Planet. Sci. Lett.* **19**, 429–37.

Nakamura N. and Matsuda H. (1989) Further characterization of fractionated and unfractionated REE and alkali metal abundances in the Allende (CV3) chondrules (II) (abstract). *14th Symposium on Antarctic Meteorites*. National Institute of Polar Research, pp. 99–100.

Neal C. R., Taylor L. A., Lui Y. and Schmitt R. A. (1991) Paired lunar meteorites MAC 88104 and MAC 88105: A new FAN of lunar petrography. *Geochim. Cosmochim. Acta* **55**, 3037–49.

Nehru C. E., Prinz M., Weisburg M. K., et al. (1992) Brachnites: A new primitive achondrite group (abstract). *Meteoritics* **27**, 267.

Nelen J., Noonan A. and Fredriksson K. (1972) Lunar glasses, breccias and chondrules. *Proc. 3rd Lunar Sci. Conf.* Lunar and Planetary Institute, pp. 723–37.

Nelson L. S., Blander M., Skaggs S. R. and Keil K. (1972) Use of a CO_2 laser to prepare chondrule-like spherules from supercooled molten oxide and silicate droplets. *Earth Planet. Sci. Lett.* **14**, 338–44.

Newsom H. E. (1995) Metal–silicate fractionation in the solar nebula (abstract). *Lunar Planet. Sci.* **XXVI**, 1043–44.

Nininger H. H. (1952) *Out of the Sky*. Dover Publications.

Noddack I. and Noddack W. (1930) Die haufigkeit der chemischen elements. *Naturwissenschaften* **18**, 757–64.

Norton O. R. (1994) *Rocks from Space*. Mountain Press Publishing Co.

Nuth J. A. III (1988) Astrophysical implications of presolar grains. In *Meteorites and the Early Solar System*. Ed. J. F. Kerridge and M. S. Matthews. University of Arizona Press, pp. 984–91.

Nyquist L., Lindstrom D., Mittlefehldt D., et al. (2001) Manganese–chromium formation intervals for chondrules from the Bishunpur and Chainpur meteorites. *Meteorit. Planet. Sci.* **36**, 911–38

Olbers H. W. M. (1803) Letter from Dr. Olbers of Bremen to Baron von Zach on the stones which have fallen from the heavens. *Phil. Mag.* **15**, 289–93.

Olmsted D. (1834) Observations of the meteors of November 13, 1833. *Amer. J. Sci.* **25**, 363–411; **36**, 137–74.

Olsen E. J. (1983) SiO_2-bearing chondrules in the Murchison (C2) meteorite. In *Chondrules and Their Origins*. Ed. E. A. King. Lunar and Planetary Institute, pp. 223–34.

Olsen E. J. and Bunch T. E. (1984) Equilibration temperatures of the ordinary chondrites – a new evaluation. *Geochim. Cosmochim. Acta* **48**, 1363–5.

Olsen E. J. and Grossman L. (1978) On the origin of isolated olivine grains in type 2 carbonaceous chondrites. *Earth Planet. Sci. Lett.* **41**, 111–27.

Olsen E. J. and Jarosewich E. (1971) Chondrules: First occurrence in an iron meteorite. *Science* **174**, 583–5.

Olsen E. J., Fredriksson K., Rajan S. and Noonan A. (1989) Chondrule-like objects and brown glasses in howardites. *Meteoritics* **25**, 187–94.

Orowan E. (1969) Density of the Moon and nucleation of planets. *Nature* **222**, 867.

Palme H. L. and Fegley B. Jr (1987) Formation of FeO-bearing olivines in carbonaceous chondrites by high temperature oxidation in the solar nebula (abstract). *Lunar Planet. Sci.* **XVIII**, 754–5.

Palme H. L., Schultz B., Spettel H. W., et al. (1981) The Acapulco meteorite: Chemistry, mineralogy, and irradiation effects. *Geochim. Cosmochim. Acta* **45**, 727–52.

Palme H. L., Larimer J. W. and Lipschutz M. E. (1988) Moderately volatile elements. In *Meteorites and the Early Solar System*. Ed. J. F. Kerridge and M. S. Matthews. University of Arizona Press, pp. 436–61.

Peck J. A. and Wood J. A. (1987) The origin of ferrous zoning in Allende chondrule olivines. *Geochim. Cosmochim. Acta* **51**, 1503–10.

Pejovic B. (1982) *Man and Meteorites*. Thomas Head.

Pellas P. (1973) Irradiation history of grain aggregates in ordinary chondrites: Possible clues to the advanced stages of accretion. In *From Plasma to Planet*. Ed. A. Elvius. John Wiley, p. 65.

Pepin R. O., Eddy J. A. and Merrill R. B., Eds. (1980) *The Ancient Sun: Fossil Record in the Earth, Moon and Meteorites; Proceedings of the Conference, Boulder, CO, October 16–19, 1979*. Pergamon Press. (*Geochim. Cosmochim. Acta*, Supplement 13.)

Pieters C. M., Taylor L. A., Noble S. K., et al. (2000) Space weathering on airless bodies: Resolving a mystery with lunar samples. *Meteorit. Planet. Sci.* **35**, 1101–7.

Podolak M. and Cameron A. G. W. (1974) Possible formation of meteoritic chondrules and inclusions in the precollapse Jovian protoplanetary atmosphere. *Icarus* **23**, 326–33.

Podolak M., Prialnik D., Bunch D. E., Cassen P. and Reynolds P. (1993) Secondary processing of chondrules and refractory inclusions (CAIs) by geodynamic heating. *Icarus* **104**, 97–107.

Podosek F. A. (1970) Dating of meteorites by the high-temperature release of iodine-correlated ^{129}Xe. *Geochim. Cosmochim. Acta* **34**, 341–65.

Podosek F. A. and Cassen P. (1994) Theoretical, observational, and isotopic estimates of the lifetime of the solar nebula. *Meteoritics* **29**, 6–25.

Podosek F. A. and Swindle T. D. (1988a) Extinct radionuclides. In *Meteorites and the Early Solar System*. Ed. J. F. Kerridge and M. S. Matthews. University of Arizona Press, pp. 1093–113.

(1988b) Nucleocosmochronology. In *Meteorites and the Early Solar System*. Ed. J. F. Kerridge and M. S. Matthews. University of Arizona Press, pp. 1114–26.

Poisson S. D. (1803) Sur les substances minérals qu l'en suppose tombées du ciel sur la terre. *Bulle. Sci. Soc. Philomat.* **3**, 180–4.

Prior G. T. (1916) The meteoritic stones of Launton, Warbreccan, Cronstad, Daniel's Kuil, Khairpur, and Soko Banja. *Mineral. Mag.* **18**, 1–25.

Proust J. L. (1805) Sur une Pierre meteorique tombée aux environs de Sigena, en Aragon, dans l'annee 1773. *Jr. Physique* **60**, 185–204; see also *Jr. Nat. Philos. (Nicholson's)* **4**, 356–7.

Radomsky P. M. and Hewins R. H. (1987) Dynamic crystallization experiments on an average type I (MgO-rich) chondrule composition. *Lunar Planet. Science* **XVIII**, 808–9.

(1988) Chondrule texture/composition relations revisited; Constraints on the thermal conditions in the chondrule forming region. *Meteoritics* **23**, 297–8.

(1990) Formation conditions of pyroxene–olivine and magnesian–olivine chondrules. *Geochim. Cosmochim. Acta* **54**, 3475–90.

Radomsky P. M., Turrin R. P. and Hewins R. H. (1986) Dynamic crystallization experiments on a pyroxene–olivine chondrule composition. *Lunar Planet. Sci.* **XVII**, 687–8.

Rambaldi E. R. (1981) Relict grains in chondrules. *Nature* **293**, 558–61.

Rambaldi E. R. and Wasson J. T. (1981) Metal and associated phases in Bishunpur, a highly unequilibrated ordinary chondrite. *Geochim. Cosmochim. Acta* **45**, 1001–15.

(1982) Fine, nickel-poor Fe–Ni grains in the olivine of unequilibrated ordinary chondrites. *Geochim. Cosmochim. Acta* **46**, 929–39.

(1984) Metal and associated phases in the highly unequilibrated ordinary chondrites Krymka and Chainpur. *Geochim. Cosmochim. Acta* **48**, 1885–97.

Rao M. N., Garrison D. H., Bogard D. D., Badhwar G. and Murali A. V. (1991) Composition of solar flare noble gases preserved in meteorite parent body regolith. *Jr. Geophys. Res.* **96**, 19321–30.

Rasmussen K. L. and Wasson I. T. (1982). A new lightning model for chondrule formation (abstract). In *Papers Presented to the Conference on Chondrules and Their Origins*. Lunar and Planetary Institute, p. 53.

Reichenbach K. L. von (1860) Meteoriten in Meteoriten. *Ann. Phys.* **111**, 353–86.

Ringwood A. E. (1959) On the evolution and densities of the planets. *Geochim. Cosmochim. Acta* **15**, 257–83.

Robert F., Javoy M., Halbout J., Dimon B. and Merlivat L. (1987) Hydrogen isotope abundances in the solar system. Part I: Unequilibrated chondrites. *Geochim. Cosmochim. Acta* **51**, 1787–805.

Robinson M. S., Thomas P. C., Veverka J., Murchie S. L. and Wilcox B. B. (2002) The geology of Eros. *Meteorit. Planet. Sci.* **37**, 1651–84.

Roedder E. (1971) Natural and laboratory crystallization of lunar glasses from Apollo 11. *Min. Soc. Japan Spec. Paper 1, Proc. IMA-IAGOD Mtg., 1970, IMA vol*, pp. 5–12.

Roedder E. and Weiblen P. W. (1977) Petrographic features and petrologic significance of melt inclusions in Apollo 14 and Apollo 15. *Proc. 8th Lunar Sci. Conf.* Lunar and Planetary Institute, pp. 2641–54.

Roedder P. L. and Emslie R. F. (1970) Olivine–liquid equilibrium. *Contrib. Mineral. Petrol.* **29**, 275–89.

Rowe P. N., Nienow A. W. and Agbim A. J. (1972) The mechanisms by which particles segregate in gas fluidised beds: Binary systems of near-spherical particles. *Trans. Inst. Chem. Engrs.* **50**, 324–33.

Rubin A. E. (1980) Kamacite and olivine in ordinary chondrites: Intergroup and intragroup relationships. *Geochim. Cosmochim. Acta* **54**, 1217–32.

(1983) The Adhi Kot breccia and implications for the origin of chondrules and silica-rich clasts in enstatite chondrites. *Earth Planet. Sci. Lett.* **64**, 201–12.

(1984a) The Blithfield meteorite and the origin of sulfide-rich, metal-poor clasts and inclusions in brecciated enstatite chondrites. *Earth Planet. Sci. Lett.* **67**, 273–83.

(1984b) Coarse-grained chondrule rims in type 3 chondrites. *Geochim. Cosmochim. Acta* **48**, 1779–89.

(1985) Impact melt products of chondritic material. *Rev. Geophys.* **23**, 277–300.

Rubin A. E. and Grossman J. N. (1987) Size–frequency distributions of EH3 chondrules. *Meteoritics* **22**, 237–51.

Rubin A. E. and Keil K. (1984) Size–distributions of chondrule types in the Inman and Allan Hills A77011 L3 chondrites. *Meteoritics* **19**, 135–43.

Rubin A. E. and Krot A. N. (1996) Multiple heating of chondrules. In *Chondrules and the Protoplanetary Disk*. Ed. R. H. Hewins, R. H. Jones and E. R. D. Scott. Cambridge University Press, pp. 173–80.

Rubin A. E., and Wasson J. T. (1986) Chondrules in the Murray CM2 meteorite and compositional differences between CM–CO and ordinary chondrite chondrules. *Geochim. Cosmochim. Acta* **50**, 307–15.

(1987a) Chondrules and matrix in the Ornans CO3 meteorite – possible precursor components. *Geochim. Cosmochim. Acta* **52**, 425–32.

(1987b) Chondrules, matrix and coarse-grained rims in the Allende meteorite: Origin. Interrelationships and possible precursor components. *Geochim. Cosmochim. Acta* **51**, 1923–37.

Rubin A. E., Scott E. R. D. and Keil K. (1982) Microchondrule-bearing clast in the Piancaldoli LL3 meteorite: A new kind of type 3 chondrite and its relevance to the history of chondrules. *Geochim. Cosmochim. Acta* **46**, 1763–76.

Rubin A. E., Fegley B. and Brett R. (1988) Oxidation state in chondrites. In *Meteorites and the Early Solar System*. Ed. J. F. Kerridge and M. S. Matthews. University of Arizona Press, pp. 488–511.

Rubin A. E., Wasson J. T., Clayton R. N. and Mayeda T. K. (1990) Oxygen isotopes in chondrules and coarse-grained chondrule rims from the Allende meteorite. *Earth Planet. Sci. Lett.* **96**, 247–55.

Russell H. N. (1929) The composition of the Sun's atmosphere. *Astrophys. Jr.* **70**, 11–82.

Russell S. S., Srinivasan G., Huss G. R., Wasserburg G. J. and McPherson G. J. (1996) Evidence for widespread ^{26}Al in the solar nebula and constraints for nebula time scales. *Science* **273**, 757–62.

Ruzicka A. (1990) Deformation and thermal histories of chondrules in the Chainpur (LL3.4) chondrite. *Meteoritics* **25**, 101–13.

Ruzicka A., Snyder G. A. and Taylor L. A. (2000) Crystal-bearing lunar spherules: Impact melting of the Moon's crust and implications for the origin of meteoritic chondrules. *Meteorit. Planet. Sci.* **35**, 173–92.

Ruzmaikina T. V. and Ip W. H. (1995) Chondrule formation in radiative shock. *Icarus* **112**, 430–47.

Ruzmaikina T. V. and Ip W. H. (1996) Chondrule formation in the accretional shock. In *Chondrules and the Protoplanetary Disk*. Ed. R. H. Hewins, R. H. Jones and E. R. D. Scott. Cambridge University Press, pp. 277–84.

Safronov V. S. (1972) *Evolution of the Protoplanetary Cloud and Formation of the Earth and the Planets*. Nauka. Translated from the Russian, *NASA Tech. Trans.*, F-677.

Sanders I. S. (1996) A chondrule-forming scenario involving molten planetesimals. In *Chondrules and the Protoplanetary Disk*. Ed. R. H. Hewins, R. H. Jones and E. R. D. Scott. Cambridge University Press, pp. 327–34.

Sauer P. N. (1993) Centrifugally driven winds from protostellar disks. I. Wind model and thermal structure. *Astrophys. J.* **408**, 115–47.

Saxena S. K. (1976) Two-pyroxene geothermometer: A model with an approximate solution. *Am. Mineral.* **61**, 643–52.

Scheeres D. J., Durda D. D. and Geissler P. E. (2002) The fate of asteroid ejecta. In *Asteroids III*. Ed. W. F. Bottke *et al*. University of Arizona Press, pp. 527–44.

Schmitt R. A., Goles G. G. and Smith R. H. (1972) Elemental abundances in stone meteorites. *Meteoritics* **7**, 131–213.

Schultz L. and Signer P. (1977) Noble gases in the St. Mesmin chondrite: Implications for the irradiation history of a brecciated meteorite. *Earth Planet. Sci. Lett.* **36**, 363–71.

Scott E. R. D. (1988) A new kind of primitive chondrite, Allan Hills 85085. *Earth Planet. Sci. Lett.* **91**, 1–18.

Scott E. R. D. and Haack H. (1993) Chemical fractionation in chondrites by aerodynamic sorting of chondritic material. *Meteoritics* **28**, 434.

Scott E. R. D. and Jones R. H. (1990) Disentangling nebula and asteroidal features of CO3 carbonaceous chondrites. *Geochim. Cosmochim. Acta* **54**, 2485–502.

Scott E. R. D. and Rajan R. S. (1981) Metallic minerals, thermal histories, and parent bodies of some xenolithic, ordinary chondrites. *Geochim. Cosmochim. Acta* **45**, 53–67.

Scott E. R. D. and Taylor G. J. (1983) Chondrules and other components in C, O, and E chondrites; Similarities in their properties and origins. *Proc. 14th Lunar Planet. Sci. Conf. J. Geophys. Res.* **88**, B275–B286.

Scott E. R. D. and Wasson J. T. (1975) Classification and properties of iron meteorites. *Rev. Geophys. Space Phys.* **13**, 527–46.

Scott E. R. D., Rubin A. E., Taylor G. J. and Keil K. (1984) Matrix material in type 3 chondrites – occurrence, heterogeneity and relationship with chondrules. *Geochim. Cosmochim. Acta* **48**, 1741–57.

Scott E. R. D., Lusby D. and Keil K. (1985) Ubiquitous brecciation after metamorphism in equilibrated ordinary chondrites. *Proc. 16th Lunar Planet. Sci. Conf. J. Geophys. Res.* **91**, E115–23.

Scott E. R. D., Barber D. J., Alexander C. M., Hutchison R. and Peck J. A. (1988) Primitive material surviving in chondrites: Matrix. In *Meteorites and the Early Solar System*. Ed. J. F. Kerridge and M. S. Matthews. University of Arizona Press, pp. 718–45.

Scott E. R. D., Love S. G. and Krot A. N. (1996) Formation of chondrules and chondrites in the protoplanetary nebula. In *Chondrules and the Protoplanetary Disk*. Ed. R. H. Hewins, R. H. Jones and E. R. D. Scott. Cambridge University Press, pp. 87–96.

Sears D. W. G. (1976) Edward Charles Howard and an early British contribution to meteorites. *J. Brit. Astron. Soc.* **86**, 133–9.

(1978a) *The Nature and Origin of Meteorites*. Adam Hilger.

(1978b) Condensation and the composition of iron meteorites. *Earth Planet. Sci. Lett.* **41**, 128–38.

(1988) *Thunderstones: The Meteorites of Arkansas*. University of Arkansas Press.

(1998) The rarity of chondrules and CAI in the early solar system and some astrophysical consequences. *Astrophys. J.* **498**, 773–8.

Sears D. W. G. and Akridge G. (1998) Nebular or parent body alteration of chondritic material: Neither or both? *Meteorit. Planet. Sci.* **33**, 1157–67.

Sears D. W. G. and Axon H. J. (1975) Metal of high cobalt content in LL chondrites. *Meteoritics* **11**, 97–100.

(1976) Nickel and cobal contents of chondritic meteorites. *Nature* **260**, 34–5.

Sears D. W. G. and Dodd R. T. (1988) Overview and classification of meteorites. In *Meteorites and the Early Solar System*. Ed. J. F. Kerridge and M. S. Matthews. University of Arizona Press, pp. 3–31.

Sears D. W. G. and Hasan F. A. (1988) Type 3 ordinary chondrites: A Review. *Surv. Geophys.* **9**, 43–97.

Sears D. W. G. and Weeks K. S. (1983) Chemical and physical studies of type 3 chondrites-II. Thermoluminescence properties of sixteen type 3 ordinary chondrites and relationships with oxygen isotopes. *Proc. 14th Lunar Planet. Sci. Conf. J. Geophy. Res.* **88**, A791–5.

Sears D. W. G., Grossman J. N., Melcher C. L., Ross L. M. and Mills A. A. (1980) Measuring metamorphic history of unequilibrated ordinary chondrites. *Nature* **287**, 791–5.

Sears D. W. G., Grossman J. N. and Melcher C. L. (1982a) Chemical and physical studies of type 3 chondrites. I; Metamorphism related studies of Antarctic and other type 3 ordinary chondrites. *Geochim. Cosmochim. Acta* **46**, 2471–81.

Sears D. W. G., Kallemeyn G. W. and Wasson J. T. (1982b) The compositional classification of chondrites: II. The enstatite chondrite groups. *Geochim. Cosmochim. Acta* **46**, 597–608.

Sears D. W. G., Sparks M. H. and Rubin A. E. (1984) Chemical and physical studies of type 3 chondrites. III: Chondrules from the Dhajala H3.8 chondrite. *Geochim. Cosmochim. Acta*, **48**, 1189–200.

Sears D. W. G., Batchelor J. D., Lu J. and Keck B. D. (1991) Metamorphism of CO and CO-like chondrites and comparisons with type 3 ordinary chondrites. *Proceedings of the NIPR Symposium Antarctic Meteorites*, volume 4. National Institute of Polar Research, pp. 319–43.

Sears D. W. G., Lu J., Benoit P. H., DeHart J. M. and Lofgren G. E. (1992) A compositional classification scheme for meteoritic chondrules. *Nature* **357**, 207–11.

Sears D. W. G., Benoit P. H. and Lu J. (1993) Two chondrule groups each with distinctive rims in Murchison recognized by cathodoluminescence. *Meteoritics* **28**, 669–75.

Sears D. W. G., Huang S. and Benoit P. H. (1995a) The formation of chondrules (abstract). *Lunar Planet. Sci.* **XXVI**, 1263–4.

Sears D. W. G., Huang S. and Benoit P. H. (1995b) Chondrule formation, metamorphism, brecciation, a new primary chondrule group, and the classification of chondrules. *Earth Planet. Sci. Lett.* **131**, 27–39.

Sears D. W. G., Huang S. and Benoit P. H. (1995c) Chondrules from the Earth and Moon: A review (abstract). *Meteoritics* **30**, 577.

Sears D. W. G., Morse A. D., Hutchison R., *et al.* (1995d) Metamorphism and aqueous alteration in low petrographic type ordinary chondrites. *Meteoritics* **30**, 169–81.

Sears D. W. G., Huang S. and Benoit P. H. (1996a) Open-system behaviour during chondrule formation. In *Chondrules and the Protoplanetary Disk*. Ed. R. H. Hewins, R. H. Jones and E. R. D. Scott. Cambridge University Press, pp. 221–31.

Sears D. W. G., Huang S., Akridge G. and Benoit P. H. (1996b) Glassy spherules in suevite from the Ries Crater, Germany, with implications for the formation of meteoritic chondrules. *Lunar Planet. Sci.* **XXVII**, 1165–6.

Sears D. W. G., Huang S., Benoit P. H., *et al.* (1997) Oxygen isotope data for classified Semarkona chondrules (abstract). *Meteorit. Planet. Sci.* **32**, A118–19.

Sears D. W. G., Lyon I., Saxton J. and Turner G. (1998) The oxygen isotopic properties of olivines in the Semarkona ordinary chondrite. *Meteorit. Planet. Sci.* **33**, 1029–32.

Sears D. W. G., Lyon I. C., Saxton J. M., Symes S. and Turner G. (1999a) Oxygen isotope heterogeneity in the mesostasis of a Semarkona group A1 chondrules. *Lunar Planet. Sci.* **XXX**, CD-ROM #1406.

Sears D. W. G., Huebner W. F. and Kochan H. W. (1999b) Laboratory simulation of the physical processes occurring on and near the surfaces of comet nuclei. *Meteorit. Planet. Sci.* **34**, 497–525.

Sears D. W. G., Allen C. C., Britt D. T., *et al.* (2002) Near-Earth Asteroid Sample Return. In *The Future of Solar System Exploration (2003–2013) – Community Contributions to the* NRC Solar System Exploration Decadal Survey (ASP *Conference Proceedings 272*). Ed. M. V. Sykes. Astronomical Society of the Pacific, pp. 111–40.

Sheng Y. J., Hutcheon I. D. and Wasserburg G. J. (1991) Origin of plagioclase–olivine inclusions in carbonaceous chondrites. *Geochim. Cosmochim. Acta* **55**, 581–99.

Shimaoka T. and Nakamura N. (1989) Vaporization of sodium from a partially molten chondritic material. *Proceedings of the NIPR Symposium on Antarctic Meteorites*, volume 2. National Institute of Polar Research, pp. 252–67.

Shu F. H., Adams F. C. and Lizano S. (1987) Star formation and molecular clouds. Observations and theory. *Ann. Rev. Astron. Astrophys.* **256**, 23–81.

Shu F. H., Najita J., Galli D., Ostriker E. and Lizano S. (1993) The collapse of clouds and the formation and evolution of stars and disks. In *Protostars and Planets III*. Ed. E. H. Levy and J. I. Lunine. University of Arizona Press, pp. 3–45.

Shu F. H., Sheng H. and Lee T. (1996) Toward an astrophysical theory of chondrites. *Science* **271**, 1545–52.

Simon S. B. and Haggerty S. E. (1980) Bulk compositions of chondrules in the Allende meteorite. *Proc. 11th Lunar Planet. Sci. Conf.* Lunar and Planetary Institute, pp. 901–27.

Skinner W. R. (1990) Bipolar outflows and a new model of the early Solar System. Part II: the origins of chondrules. *Lunar Planet. Sci.* **XXI**, 1168–9.

Skinner W. R. and Leenhouts J. H. (1993) The size distribution and aerodynamic equivalence of metal chondrules and silicate chondrules in Acfer 059. *Lunar Planet. Sci.* **XXIV**, 1315–16.

Smales A. A., Mapper D. and Wood A. J. (1957) The determination by radioactiviation of small quantities of nickel, cobalt, and copper in rocks, marine sediments and meteorites. *Analyst* **82**, 75.

Smith J. V. (1982) Heterogeneous growth of meteorites and planets, especially the earth and moon. *J. Geol.* **90**, 1–48.

Sonett C. P. (1979) On the origin of chondrules. *Geophys. Res. Lett.* **6**, 677–80.

Sorby H. C. (1864) On the microscopical structure of meteorites. *Phil. Mag.* **28**, 157–9.
  (1877) On the structure and origin of meteorites. *Nature* **15**, 495–8.

Space Studies Board (2002) *New Frontiers in the Solar System: An Integrated Exploration Strategy*. National Research Council.

Srinivasan G., Huss G. R. and Wasserburg G. J. (2000) A petrographic, chemical, and isotopic study of calcium–aluminum-rich inclusions and aluminum-rich chondrules from the Axtell (CV3) chondrite. *Meteorit. Planet. Sci.* **35**, 1333–54.

Steele I. M. (1985) Compositions and textures of relic forsterite in carbonaceous and unequilibrated ordinary chondrites. *Geochim. Cosmochim. Acta* **50**, 1379–95.
  (1988) Primitive material surviving in chondrites: Mineral grains. In *Meteorites and the Early Solar System*. Ed. J. F. Kerridge and M. S. Matthews. University of Arizona Press, pp. 808–18.

Stepinski T. F. and Reyes-Ruiz M. (1993) Magnetically controlled solar nebula. *Lunar Planet. Sci.* **XXIV**, 1351–2.

Stöffler D., Bischoff A., Buchwald V. and Rubin A. E. (1988) Shocke effects in meteorites. In *Meteorites and the Early Solar System*. Ed. J. F. Kerridge and M. S. Matthews. University of Arizona Press, pp. 165–202.

Suess H. E. and Thompson W. B. (1983) Can chondrules form from a gas of solar composition? In *Chondrules and Their Origins*. Ed. E. A. King. Lunar and Planetary Institute, pp. 243–5.

Sugiura N. and Strangway D. W. (1985) NRM directions around a centimeter sized dark inclusion in Allende. *Proc. 15th Lunar Planet. Sci. Conf. J. Geophys. Res.* **90**, C729–38.
  (1988) Magnetic studies of meteorites. In *Meteorites and the Early Solar System*. Ed. J. F. Kerridge and M. S. Matthews. University of Arizona Press, pp. 595–615.

Sugiura N., Lanoix, M. and Strangway D. W. (1979) Magnetic fields of the solar nebula as recorded in chondrules from the Allende meteorite. *Phys. Earth Planet. Int.* **20**, 342–9.

Sullivan R., Grelley R., Pappalardo R., et al. (1996) Geology of 243 Ida. *Icarus* **142**, 89–96.

Swindle T. D. (1988) Trapped noble gases in meteorites. In *Meteorites and the Early Solar System*. Ed. J. F. Kerridge and M. S. Matthews. University of Arizona Press, pp. 535–64.

Swindle T. D. and Grossman J. N. (1987) I–Xe studies of Semarkona Chondrules: Dating alteration (abstract). *Lunar Planet. Sci.* **XVIII**, 982–3.

Swindle T. D. and Podosek F. A. (1988) Iodine–xenon dating. In *Meteorites and the Early Solar System*. Ed. J. F. Kerridge and M. S. Matthews. University of Arizona Press, pp. 1127–46.

Swindle T. D., Caffee M. W. and Hohenberg C. M. (1983a) Radiometric ages of chondrules. In *Chondrules and Their Origins*. Ed. E. A. King. Lunar and Planetary Institute, pp. 246–61.

Swindle T. D., Caffee M. W., Hohenberg C. M. and Lindstrom M. M. (1983b) I–Xe studies of individual Allende Chondrules. *Geochim. Cosmochim. Acta* **47**, 2157–77.

Swindle T. D., Caffee M. W. and Hohenberg C. M. (1986) I–Xe and ^{40}Ar–^{39}Ar ages of Chainpur chondrules (abstract). *Lunar Planet. Sci.* **XVII**, 857–8.

Swindle T. D., Grossman J. N., Olinger C. T. and Garrison D. H. (1991) Iodine–xenon, chemical, and petrographic studies of Semarkona chondrules – Evidence for the timing of aqueous alteration. *Geochim. Cosmochim. Acta* **55**, 3723–34.

Swindle T. D., Davis A. M., Hohenberg C. M., MacPherson G. J. and Nyquist L. E. (1996) Formation times of chondrules and Ca–Al-rich inclusions: Constraints from short-lived radionuclides. In *Chondrules and the Protoplanetary Disk*. Ed. R. H. Hewins, R. H. Jones and E. R. D. Scott. Cambridge University Press, pp. 77–86.

Symes S. J. K., Sears D. W. G., Akridge D. G., Huang S. and Benoit P. H. (1998) The crystalline lunar spherules: Their formation and implications for the origin of meteoritic chondrules. *Meteorit. Planet. Sci.* **33**, 13–29.

Takahashi H., Janssens M. J-., Morgan J. W. and Anders E. (1978a) Further studies of trace elements in C3 chondrites. *Geochim. Cosmochim. Acta* **42**, 97–107.

Takahashi H., Gros J., Higuchi H., Morgan J. W. and Anders E. (1978b) Volatile elements in chondrites: metamorphism or nebular? *Geochim. Cosmochim. Acta* **42**, 1859–69.

Tatsumoto M., Unmh D. M. and Desborough G. A. (1976) U–Th–Pb and Rb–Sr systematics of Allende and U–Th–Pb systematics of Orgueil. *Geochim. Cosmochim. Acta* **40**, 617–34.

Taylor G. J., Scott E. R. D. and Keil K. (1983) Cosmic setting for chondrule formation. In *Chondrules and Their Origins*. Ed. E. A. King. Lunar and Planetary Institute, pp. 262–78.

Taylor G. J., Scott E. R. D., Keil K., *et al.* (1984) Primitive nature of ordinary chondrite matrix materials. *Lunar Planet. Sci.* **XV**, 848–9.

Taylor G. J., Maggiore P., Scott E. R. D., Rubin A. E. and Keil K. (1987) Original structures, and fragmentation and reassembly histories of asteroids: Evidence from meteorites. *Icarus* **69**, 1–13.

Taylor L. A. and Cirlin E. H. (1986) Olivine/melt Fe/Mg K_d's <0.3: Rapid cooling of olivine-rich chondrules. *Lunar Planet. Sci.* **XVII**, 879–80.

Thiemens M. H. (1988) Heterogeneity in the nebula: Evidence from stable isotopes. In *Meteorites and the Early Solar System*. Ed. J. F. Kerridge and M. S. Matthews. University of Arizona Press, pp. 899–923.

  (1996) Mass-independent isotopic effects in chondrites: The role of chemical processes. In *Chondrules and the Protoplanetary Disk*. Ed. R. H. Hewins, R. H. Jones and E. R. D. Scott. Cambridge University Press, pp. 107–18.

Thomas P., Adinolfi D., Helfenstein P., Simonelli D. and Veverka J. (1996) The surface of Diemos: Contribution of materials and processes to its unique appearance. *Icarus* **123**, 536–56.

Thomas P. C., Veverka J., Bell J. F., *et al.* (1999) Mathilde, size, shape and geology. *Icarus* **140**, 17–27.

Thomas P. C., Veverka J., Sullivan R., *et al.* (2000) Phobos: Regolith and ejecta blocks investigated with Mars orbiter camera images. *J. Geophys. Res.* **105**, 15091–106.

Thomas P. C., Joseph J., Robinson M., *et al.* (2002) Shape, slopes, and slope processes on Eros. *Icarus* **155**, 18–37.

Tilton G. R. (1988a) Age of the solar system. In *Meteorites and the Early Solar System*. Ed. J. F. Kerridge and M. S. Matthews. University of Arizona Press, pp. 259–75.

  (1988b) Principles of radiometric dating. In *Meteorites and the Early Solar System*. Ed. J. F. Kerridge and M. S. Matthews. University of Arizona Press, pp. 249–58.

Tissandier L., Libourel G. and Robert F. (2002) Gas–melt interactions and their bearing on chondrule formation. *Meteorit. Planet. Sci.* **37**, 1377–89.

Tomeoka K. (1990) Phyllosilicate veins in the Yamato-82162 CI carbonaceous chondrite: Evidence for post-accretionary aqueous alteration (abstract). *Meteoritics* **25**, 415.

Tomeoka K. and Buseck P. R. (1982) Intergrown mica and montmorillonite in the Allende carbonaceous chondrite. *Nature* **299**, 326–7.

Tomeoka K., McSween H. Y. and Buseck P. R. (1989) Mineralogical alteration of CM carbonaceous chondrites: A review. *Proceedings of the NIPR Symposium on Antarctic Meteorites*, volume 2. National Institute of Polar Research, pp. 221–34.

Trieloff M., Jessberger E. K., Herrwerth I., *et al.* (2003) Structure and thermal history of the H-chondrite parent asteroid revealed by thermochronometry. *Nature* **422**, 502–6.

Tschermak G. (1883) Beitrag zur Classification der Meteoriten. *Sitzber. Akad. Wiss. Wien, Math. -Naturw. Cl.* **85** (1), 347–71.

(1885) Die mikroskopische Beschaffenheit der Meteoriten. *Smithson. Contrib. Astrophys.* **4**, 138–234 (1964, translated by J. A. Wood and E. M. Wood).

Tsuchiyama A. and Nagahara H. (1981) Effects of precooling thermal history and cooling rate on the texture of chondrules; A preliminary report. *Mem. Natl Inst. Polar Res., Special Issue* **20**, 175–92.

Tsuchiyama A., Nagahara H. and Kushiro I. (1980) Experimental reproduction of textures of chondrules. *Earth Planet. Sci. Lett.* **48**, 155–65.

Tsuchiyama A., Nagahara H. and Kushiro I. (1981) Volatilization of sodium from silicate melt spheres and its application to the formation of chondrules. *Geochim. Cosmochim. Acta*, **45**, 1357–67.

Turner G. (1988) Dating of secondary events. In *Meteorites and the Early Solar System*. Ed. J. F. Kerridge and M. S. Matthews. University of Arizona Press, pp. 276–88.

Urey H. C. (1952) *The Planets*. Yale University Press.

(1956) Diamonds, meteorites, and the origin of the solar system. *Astrophys. J.* **124**, 623–37.

(1958) The early history of the solar system as indicated by the meteorites. *Proc. Chem. Soc. (March)*, 67–78.

(1961) Criticism of Dr. B. Mason's paper on the 'The Origin of Meteorites'. *J. Geophys. Res.* **66**, 1988–91.

(1962) Evidence regarding the origin of the earth. *Geochim. Cosmochim. Acta* **26**, 1–13.

(1967) Parent bodies of meteorites and the origin of chondrules. *Icarus* **7**, 350–9.

Urey H. C. and Craig H. (1953) The composition of the stone meteorites and the origin of the meteorites. *Geochim. Cosmochim. Acta* **4**, 36–82.

Urey H. C. and Donn B. (1956) Chemical heating for meteorites. *Astrophys. J.* **124**, 307–10.

Valentine G. A. and Fisher R. V. (1993) Glowing avalanches: New research on volcanic density currents. *Science* **259**, 1130–1.

Van Schmus W. R. (1969) The mineralogy and petrology of chondritic meteorites. *Earth Sci. Rev.* **5**, 145–84.

Van Schmus W. R. and Hayes J. M. (1974) Chemical and petrographic correlations among carbonaceous chondrites, *Geochim. Cosmochim. Acta* **38**, 47–64.

Van Schmus W. R. and Wood J. A. (1967) A chemical–petrologic classification for the chondritic meteorites. *Geochim. Cosmochim. Acta* **31**, 747–65.

Veverka J. and Duxbury T. C. (1977) Viking observations of Phobos and Diemos: Preliminary results. *J. Geophys. Res.* **82**, 4213–23.

Veverka J. and Thomas, P. (1979) Phobos and Deimos: A preview of what asteroids are like. In *Asteroids*. Ed. T. Gehrels. University of Arizona Press, pp. 628–51.

Veverka J., Robinson M., Thomas P., *et al.* (2000) NEAR at Eros: imaging and spectral results. *Science* **289**, 2088–97.

Veverka J., Farquhar B., Robinson M., *et al.* (2001) The landing of the NEAR-Shoemaker spacecraft on asteroid 433 Eros. *Nature* **413**, 390–3.

Vilas F. (1994) A cheaper, faster, better way to detect water of hydration on Solar System bodies. *Icarus* **111**, 456–67.

Von Michaelis H., Willis J. P., Erlank A. J. and Ahrens L. H. (1969a) The composition of stony meteorites I. Analytical techniques. *Earth Planet. Sci. Lett.* **5**, 383–6.

Von Michaelis H., Ahrens L. H. and Willis J. P. (1969b) The composition of stony meteorites II. The analytical data and an assessmant of their quality. *Earth Planet. Sci. Lett.* **5**, 387–94.

Wahl W. A. (1910a) Beiträge zur Chemie der Meteoriten. *Z. Anorgan. Chem.* **69**, 52–96.

Wahl W. A. (1910b) The brecciated stony meteorites and meteorites containing foreign fragments. *Geochim. Cosmochim. Acta* **2**, 91–117.

Wai C. M. and Wasson J. T. (1977) Nebular condensation of moderately volatile elements and their abundances in ordinary chondrites. *Earth Planet. Sci. Lett.* **36**, 1–13.

Walter L. S. and Dodd R. T. (1972) Evidence for vapor fractionation in the origin of chondrules. *Meteoritics* **7**, 341–52.

Wark D. A. and Lovering J. F. (1982) Evolution of Ca–Al-rich bodies in the earliest solar system: Growth by incorporation. *Geochim. Cosmochim. Acta* **46**, 2595–607.

Wark D. A., Boynton W. V., Keays R. R. and Palme H. (1987) Trace element clues to the formation of forsterite-bearing inclusions in the Allende meteorite. *Geochim. Cosmochim. Acta* **51**, 607–22.

Warren P. H., Jerde E. A. and Kallemeyn G. W. (1990) Pristine moon rocks: An alkali anorthosite with coarse augite exsolution from plagioclase, a magnesian harzburgite, and other oddities. *Proc. 20th Lunar Planet. Sci. Conf.* Lunar and Planetary Institute, pp. 2641–54.

Wasserburg G. J. (1985) Short-lived nuclei in the early solar system. In *Protostars and Planets II*. Ed. D. C. Black and M. S. Matthews. University of Arizona Press, pp. 703–37.

Wasson J. T. (1972) Formation of ordinary chondrites. *Rev. Geophys. Space Phys.* **10**, 711–59.

  (1974) *Meteorites*. Springer-Verlag.

  (1985) *Meteorites: Their Record of Early Solar-System History*. W. H. Freeman.

  (1993) Constraints on chondrule origins. *Meteoritics* **28**, 14–28.

  (1996) Chondrule formation: Energetics and length scales. In *Chondrules and the Protoplanetary Disk*. Ed. R. H. Hewins, R. H. Jones and E. R. D. Scott. Cambridge University Press, pp. 45–54.

Wasson J. T. and Chou C. L. (1974) Fraction of moderately volatile elements in ordinary chondrites. *Meteoritics* **9**, 69–84.

Wasson J. T. and Rasmussen K. L. (1994) The fine nebula dust component: A key to chondrule formation by lightning (abstract). *Papers Presented to the Conference on Chondrules and the Protoplanetary Disk*. 43. Lunar and Planetary Institute.

Watanabe S., Kitamura M. and Morimoto N. (1984) Analytical electron microscopy of a chondrule with relict olivine in the ALH-77015 chondrite (L3). *Mem. Natl Inst. Polar Res., Special Issue* **35**, 200–9.

  (1986) Oscillatory zoning of pyroxenes in ALH-77214 (L3) (abstract). *Proceedings of 11th Symposium on Antarctic Meteorites*. National Institute of Polar Research, pp. 25–7.

Wdowiak T. J. (1983) Experimental investigation of electrical discharge formation of chondrules. In *Chondrules and Their Origins*. Ed. E. A. King. Lunar and Planetary Institute, pp. 279–83.

Weidenschilling S. J. (1977) Aerodynamics of solid bodies in the solar nebula. *Mon. Not. Roy. Astron. Soc.* **180**, 57–70.

(1988) Formation processes and time scales for meteorite parent bodies. In *Meteorites and the Early Solar System*. Ed. J. F. Kerridge and M. S. Matthews. University of Arizona Press, pp. 348–71.

Weidenschilling S. J. and Ruzmaikina T V. (1994) Coagulation of grains in static and collapsing protostellar clouds. *Astrophys. J.* **430**, 713–26.

Weidenschilling S. J., Marzari F. and Hood L. L. (1998) The origin of chondrules at Jovian resonances. *Science* **279**, 681–4.

Weinbruch S., Buettner H., Holzheid A., Rosenhauer M. and Hewins R. H. (1998) On the lower limit of chondrule cooling rates: The significance of iron loss in dynamic crystallization experiments. *Meteorit. Planet. Sci.* **33**, 65–74.

Weinbruch S., Müller W. F. and Hewins R. H. (2001) A transmission electron microscope study of exsolution and coarsening in iron-bearing clinopyroxene from synthetic analogues of chondrules. *Meteorit. Planet. Sci.* **36**, 1237–48.

Weisberg M. K. and Prinz M. (1996) Agglomeratic chondrules, chondrule precursors, and incomplete melting. In *Chondrules and the Protoplanetary Disk*. Ed. R. H. Hewins, R. H. Jones and E. R. D. Scott. Cambridge University Press, pp. 119–27.

Weisberg M. K., Nehru C. E. and Prinz M. (1988) Petrology of ALH85085 – a chondrite with unique characteristics. *Earth Planet. Sci. Lett.* **91**, 19–32.

Weisberg M. K., Prinz M., Kojima H., et al. (1991) The Carlisle Lake-type chondrites: A new grouplet with high $\delta^{17}O$ and evidence for nebular oxidation. *Geochim. Cosmochim. Acta* **55**, 2657–69.

Weisberg M. K., Prinz M., Clayton R. N. and Mayeda T. (1993) The CR (Renazzo-type) carbonaceous chondrite group and its implications. *Geochim. Cosmochim. Acta* **55**, 2657–69.

Weisberg M. K., Prinz M., Clayton R. N., et al. (1996) The K (Kakangari) chondrite grouplet. *Geochim. Cosmochim. Acta* **60**, 4253–63.

Wetherill G. W. (1985) Asteroidal source of ordinary chondrites. *Meteoritics* **20**, 1–22.

Wetherill G. W. and Chapman C. R. (1988) Asteroids and meteorites. In *Meteorites and the Early Solar System*. Ed. J. F. Kerridge and M. S. Matthews. University of Arizona Press, pp. 35–67.

Whipple F. L. (1966) Chondrules: Suggestions concerning their origin. *Science* **153**, 54–6.

(1972a) Accumulation of chondrules on asteroids. In *Physical Studies of Minor Planets*. NASA Special Publication 267, pp. 251–62.

(1972b) On certain aerodynamic processes for asteroids and comets. In *From Plasma to Planet, Nobel Symposium 21*. Ed. A. Elvius. John Wiley, pp. 211–32.

Wieneke B. and Clayton D. D. (1983) Aggregation of grains in a turbulent pre-solar disk. In *Chondrules and Their Origins*. Ed. E. A. King. Lunar and Planetary Institute, pp. 284–95.

Wiik H. B. (1969) On regular discontinuities in the composition of meteorites. *Comm. Phys. Math.* **34**, 135–45.

Wilkening L. L., Boynton W. V. and Hill D. H. (1984) Trace elements in rims and interiors of Chainpur chondrules. *Geochim. Cosmochim. Acta* **48**, 1071–80.

Wilson C. J. N. (1980) The role of fluidization in the emplacement of pyroclastic flows: An experimental approach. *J. Volcanol. Geotherm. Res.* **8**, 231–49.

Winzer S. R., Nava D. F., Meyerhoff M., et al. (1977) The petrology and geochemistry of impact melts, granulites, and hornfeldses from consortium breccia 61175. *Proc. 8th Lunar Sci. Conf.* Lunar and Planetary Institute, pp. 1943–66.

Wisdom J. (1985) Meteorites follow a chaotic route to Earth. *Nature* **315**, 731–3.

Wlotzka F. (1969) On the formation of chondrules and metal particles by shock melting. In *Meteorite Research*. Ed. P. M. Millman. D. Reidel, pp. 174–83.

Wlotzka F. (1983) Composition of chondrules, fragments and matrix in the unequilibrated ordinary chondrites Tieschitz and Sharps. In *Chondrules and Their Origins*. Ed. E. A. King. Lunar and Planetary Institute, pp. 296–318.

Wolf R., Richter G. R., Woodrow A. B. and Anders E. (1980) Chemical fractionations in meteorites – XI. C2 chondrites. *Geochim. Cosmochim. Acta* **44**, 711–17.

Wood J. A. (1962) Metamorphism in chondrites. *Geochim. Cosmochim. Acta* **26**, 739–49.

(1963) The origin of chondrules and chondrites. *Icarus* **2**, 152–80.

(1964) The cooling rates and parent planets of several iron meteorites. *Icarus* **3**, 429–59.

(1967a) Chondrites: Their metallic minerals, thermal histories and parent planets. *Icarus* **6**, 1–7.

(1967b) Olivine and pyroxene compositions in type II carbonaceous chondrites. *Geochim. Cosmochim. Acta* **31**, 2095–108.

(1968) *Meteorites and the Origin of the Planets*. McGraw-Hill.

(1979) Review of metallographic cooling rates of meteorites and a new model for the planetesimals in which they formed. In *Asteroids*. Ed. T. Gehrels. University of Arizona Press, pp. 849–91.

(1983) Formation of chondrules and CAI's from interstellar grains accreting to the solar nebula. *Mem. Natl. Inst. Polar Res., Special Issue* **30**, 84–92.

(1984) On the formation of meteoritic chondrules by aerodynamic drag heating in the solar nebula. *Earth Planet. Sci. Lett.* **70**, 11–26.

(1985) Meteoritic constraints on processes in the solar nebula. In *Protostars and Planets II*. Ed. D. C. Black and M. S. Matthews. University of Arizona Press, pp. 687–702.

(1986) High temperatures and chondrule formation in a turbulent shear zone beneath the nebula surface. *Lunar Planet. Science* **XVII**, 456–957.

(1988) Chondritic meteorites and the solar nebula. *Ann. Rev. Earth Planet. Sci.* **16**, 53–72.

(1996) Unresolved issues in the formation of chondrules and chondrites. In *Chondrules and the Protoplanetary Disk*. Ed. R. H. Hewins, R. H. Jones and E. R. D. Scott. Cambridge University Press, pp. 55–69.

(2001) *Chondrites: Tight-lipped witnesses to the beginning*. Unpublished article widely distributed by author.

Wood J. A. and Chang S., Eds. (1985) *The Cosmic History of the Biogenic Elements and Compounds*. NASA SP-476.

Wood J. A. and Hashimoto A. (1988) The condensation sequence under non-classic conditions ($P < 10^{-3}$ atm, non-cosmic compositions). *Lunar Planet. Sci* **XIX**, 1292–3.

(1993) Mineral equilibrium in fractionated nebular systems. *Geochim. Cosmochim. Acta* **57**, 2377–88.

Wood J. A. and McSween H. Y. Jr (1976) Chondrules as condensation products. In *Comets, Asteroids, Meteorites: Interrelations, Evolution, and Origins*. Ed. A. H. Delsemme. University of Toledo, pp. 65–373.

Wood J. A. and Morfill G. E. (1988) A review of solar nebula models. In *Meteorites and the Early Solar System*. Ed. J. F. Kerridge and M. S. Matthews. University of Arizona Press, pp. 329–47.

Woolum D. S. (1988) Solar system abundances and processes of nucleosynthesis. In *Meteorites and the Early Solar System*. Ed. J. F. Kerridge and M. S. Matthews. University of Arizona Press, pp. 995–1020.

Young E. D. and Russell S. S. (1998) Oxygen reservoirs in the early solar nebula inferred from an Allende CAI. *Science* **282**, 452–5.

Young J. (1926) The crystal structure of meteoric iron as determined by X-ray analysis. *Proc. Roy. Soc. London* **A112**, 630–41.

Yu Y., Hewins R. H., Clayton R. N. and Mayeda T. K. (1995) Experimental study of high temperature oxygen isotope exchange during chondrule formation. *Geochim. Cosmochim. Acta* **59**, 2095–104.

Yu Y., Hewins R. H. and Zanda B. (1996) Sodium and sulfur in chondrules: Heating time and cooling curves. In *Chondrules and the Protoplanetary Disk*. Ed. R. H. Hewins, R. H. Jones and E. R. D. Scott. Cambridge University Press, pp. 213–19.

Zbik M. and Lang B. (1983) Morphological features of pore spaces in chondrules. In *Chondrules and Their Origins*. Ed. E. A. King. Lunar and Planetary Institute, pp. 319–29.

Zhang Y., Benoit P. H. and Sears D. W. G. (1995) The classification and complex thermal history of the enstatite chondrites. *J. Geophys. Res.* **100**, 9417–38.

Zinner E. (1988) Interstellar cloud material in meteorites. In *Meteorites and the Early Solar System*. Ed. J. F. Kerridge and M. S. Matthews. University of Arizona Press, pp. 956–83.

Zolensky M. and McSween H. Y. Jr (1988) Aqueous alteration. In *Meteorites and the Early Solar System*. Ed. J. F. Kerridge and M. S. Matthews. University of Arizona Press, pp. 114–43.

Zook H. A. (1980) A new impact theory for the generation of ordinary chondrites. *Meteoritics* **15**, 390–1.

  (1981) On a new theory for the generation of chondrules. *Lunar Planet. Sci.* **XII**, 1242–4.

Index

acapulcoites 50
accretion 35, 57, 60, 61, 88, 95, 98, 111, 125, 137, 138, 142, 144, 145, 157
achondrites 14, 27, 50, 66
  primitive 14, 50
ages 32, 33, 34, 39, 61, 63, 65, 67–70, 94, 104, 130, 132, 137
  Ar–Ar 32, 34, 67, 70, 132
  chondrules 1, 2, 3, 5, 6, 10, 12, 15, 16, 17, 20, 35–37, 47, 50, 53, 55, 57, 58, 60, 63, 64, 70, 71, 73–120, 122, 123
  compaction 61
  cosmic ray exposure 32, 33, 67, 70
  H chondrites 9, 24, 25, 29, 32, 33, 39, 70, 132, 145
  I–Xe 67, 68, 94
  K–Ar 34, 67
  Pb–Pb 63, 64
  Rb–Sr 9, 10
  U–Th 64
agglomeration 60
agglutinates 31, 44, 117, 123, 130, 140
airbursts 131
albedo 23, 25, 47
aluminum-26 67
  in chondrules 36, 75, 80, 83, 85, 90, 92, 97, 102, 103, 105, 130, 133
  in refractory inclusions 67, 128
angular momentum 126
Antarctica 3, 11, 12
Apollo 10, 12, 19, 30, 46, 106, 109, 136, 140
aqueous alteration 15, 60, 63, 64, 65, 88, 94, 103
asteroid 23, 25, 47
  albedo 25, 47
  C class 25
  classes 6, 7, 8, 11, 14, 15, 16, 18, 20, 25, 28, 29, 30, 34, 49, 50, 54, 55, 58, 59, 63, 70, 73, 77, 80, 82, 83, 85, 93, 96, 105, 106, 107, 109, 115, 127, 129, 132, 133, 139, 141, 146, 151, 153, 157, 158
  density 21, 22, 23, 24, 26, 43, 63, 115, 117, 118, 114, 131, 133, 136, 143, 145, 147, 148, 151, 154, 156
  ejecta 31, 36, 43, 44, 45, 47, 109, 113, 122, 138, 139, 140, 160, 161
  formation 10, 13, 16, 20, 31, 34, 47, 55, 57, 58, 60, 66, 67–70, 73, 82–90, 93–95, 98, 100, 102–106, 111–114, 117, 119–139, 143, 147, 148, 151, 152, 154, 157, 159
  geology 4, 42, 137, 160
  heat source 37, 38, 39, 67, 129
  impact 10, 11, 15, 20, 25, 31, 35, 37, 38, 39, 41, 43, 44, 46, 47, 60, 61, 88, 109, 110, 111, 112, 113, 117, 119, 120, 122, 123, 129, 130, 134, 135, 137, 138–140, 143, 148, 151, 152, 153, 155–159
  internal structure 139
  magnitude 35, 47, 53, 85, 115
  Mars-orbiters 42
  meteorite links 29
  near-Earth 11, 21, 29, 43, 160, 161, 162
  orbits 24, 25, 31, 154
  regolith properties 45
  rotation rates 47
  S class 29
  sample return 11, 160, 161
  size 11, 14, 16, 31, 32, 33, 35, 39, 44, 45, 46, 47, 58, 88, 89, 90, 95, 96, 98, 101, 103, 109, 110, 116, 133, 138, 140, 141, 142, 143, 146, 147, 148, 151, 152, 154, 156, 157, 158
  spectra 21, 25, 26, 28, 29, 31, 40, 132, 159, 160
asteroids, named 11, 21
  1998 SF36 22
  1999 KW4 22
  2000 DP107 22
  2000 UG11 22
  Annefrank 43
  Antiope 22
  Braille 43
  Ceres 22
  Eros 11, 21, 22, 40, 41, 42, 43, 138
  Eugenia 22
  Eunomia 22
  Gaspra 40, 41, 43, 45, 138, 140
  Hebe 24, 25, 29, 36
  Hermione 22
  Hispania 22
  Hygeia 22
  Ida 22, 31, 40, 42, 43, 81, 138, 140

asteroids, named (*cont.*)
  Interamnia 22
  Kalliope 22
  Massalia 22
  Mathilde 11, 22, 40, 41, 42, 43, 138, 140
  Pallas 22
  Parthenope 22
  Psyche 22, 41, 43
  Pulcova 22
  Sylvia 22
  Vesta 22, 26, 29, 66
astrophysics 6, 112
aubrites 27, 67, 68
  I–Xe ages 68, 94
Australia 112

BABI 66
Berzelius, Jons Jacob 3
bipolar outflows 111, 114, 125, 126
Bournon, Jacques Louis Compte de 1, 2, 129
brachinites 27, 50
breccias 19, 20, 49, 54, 55, 70, 71, 141, 143, 145, 147, 149, 152, 156, 157, 158
  fragmental 61, 97, 121
  granultic 61
  impact-melt 47, 109, 110, 120, 122, 123, 130, 138, 152, 156
  lithic 61, 75, 77, 82, 105, 143
  lunar 3, 4, 10, 11, 31, 32, 90, 109, 117, 122, 130, 136, 139, 140, 143, 159
  primitive 3, 8, 12, 14, 50, 60, 61, 70, 79, 80, 105, 117, 122, 126, 132–134, 152, 155, 156, 158, 159
  regolith 11, 19, 30, 31, 32, 35, 41, 43–45, 47, 61, 109, 117, 123, 130, 137, 138, 139, 140, 147, 151–153, 158, 160
brecciation 20, 39, 60, 61, 122, 137, 153

C asteroids 21, 25, 26, 29, 69, 156
  abundance 7, 8, 9, 14, 15, 27, 33, 53, 54, 55, 57, 58, 60, 68, 73, 80, 82, 83, 88–90, 104, 109, 128, 129, 143, 152, 156, 157
C chondrites 21, 26, 29, 49, 54
  I–Xe ages 68, 94
Ca/Si 14, 106, 127
calcium–aluminum inclusions (refractory inclusions) 1, 16, 80, 106
calculations 22, 36, 37, 39, 44, 45, 54, 55, 56, 57, 107, 109, 110, 114, 131, 138, 140, 142, 143, 145
  ballistic 109, 110, 140, 141–143
  ejecta 31, 36, 43, 44, 45, 47, 109, 113, 122, 138, 139, 140, 160, 161
  metal–silicate segregation 7, 20, 49, 54, 55, 70, 71, 141, 143, 145, 147, 149, 152, 156, 157, 158
  thermal models 35, 36, 123
  thermodynamic 54, 55, 56, 57, 85
carbon 14, 31, 88, 90, 121, 156, 157, 158
carbon dioxide 90
carbon monoxide 19, 49, 54, 63, 98, 106, 157
carbonaceous chondrites 25, 49, 58, 59, 61, 64, 77, 81, 97, 98, 99, 105, 109, 127, 155, 156, 157
  atmospheric passage 131, 132
  formation
  oxygen isotopes 101, 153, 158

cathodoluminescence 79, 80, 100, 105, 123
CCAM line 103, 157
CH chondrites 49, 158
charged particle tracks 29, 30, 61, 69, 135, 136
Chladni, Ernst Florins 2
chondrite ages 70, 130
chondrite classes 6, 7, 8, 11, 14, 15, 16, 18, 20, 30, 49, 50, 54, 55, 58, 59, 63, 70, 77, 82, 85, 105, 106, 115, 127, 133, 146, 151, 153, 157, 158
  CH 14, 49, 75, 107, 158
  CI 79, 100, 123
  CK 14, 49
  CM 3, 5, 8, 14, 16, 19, 23, 49, 54, 58, 61, 63, 64, 65, 68, 96, 97, 105, 106, 109, 115, 117, 131, 133, 140, 156, 157
  CO 14, 16, 19, 49, 50, 51, 54, 63, 68, 81, 83, 84, 98, 99, 105, 106, 145, 156, 157
  CR 14, 49, 54, 82, 83, 84, 107, 114, 146
  CV 14, 16, 19, 49, 54, 58, 63, 68, 99, 101, 103, 105, 106, 107, 156, 157
  EH 14, 16, 49, 53, 54, 63, 70, 105, 156, 158
  EL 14, 16, 49, 53, 54, 63, 70, 105, 115, 156, 158
  enstatite 16, 20, 27, 49, 53, 56, 57, 58, 59, 81, 105, 106, 114, 118, 123, 127, 157, 158
  H 3, 4, 6, 9, 14, 16, 24, 25, 27, 29, 30, 31, 32, 35, 39, 41, 47, 53, 54, 55, 56, 57, 59, 67, 68, 70, 75, 77, 90, 98, 105, 113, 122, 132, 145, 146, 148, 156, 161
  L 6, 14, 16, 25, 27, 32, 33, 34, 49, 53, 54, 55, 59, 67, 68, 70, 77, 98, 105, 116, 122, 132, 142, 145, 146, 148, 156
  LL 14, 16, 21, 27, 32, 33, 49, 53, 54, 55, 59, 67, 68, 70, 98, 105, 132, 146, 148, 156
  R 14, 23, 27, 44, 49, 96, 102, 115, 119
chondrite components 133
  calcium–aluminum inclusions (CAI) 16
  chondrules 1–3, 5, 6, 10, 12, 15, 16, 17, 20, 35, 36, 37, 47, 50, 53, 55, 57, 58, 60, 63, 64, 70–73, 75, 77, 79–95, 105, 106, 107, 109–112, 120–123, 126–135, 142, 143, 146–159
  matrix 1, 3, 5, 15, 16, 31, 49, 55, 58, 60, 64, 151, 152, 156
  metal 6, 7, 15, 20, 29, 32, 35, 37, 39, 43, 49–55, 57, 58, 61, 70, 71, 73, 77, 82, 85, 88, 94, 97, 107, 115, 121–123, 127, 130, 134, 135, 138, 141, 143–147, 149, 151, 152, 153, 156–159
  refractory inclusions 15, 32, 49, 51, 55, 57, 61, 73, 82, 94, 97, 134
  sulfide 65, 66, 67, 69, 101, 105, 106, 107, 128, 129
chondrite composition 31, 134, 151
  INAA 50, 80
  isotope dilution 8, 50
  noble gases 51, 164, 188
  oxygen isotopes 63, 101, 153
  X-ray fluorescence 7, 50, 64
chondrite history 10, 11, 58, 63, 98
  thermal models 35, 123
chondrite origin 10, 20, 71, 111, 141
  parent bodies 10, 20, 21, 30, 32, 35, 39, 40, 70, 95, 119, 123, 132, 138, 151
  parent body structures 39

chondrite properties 127, 128, 141
  Ca/Si 106, 127
  carbon 25, 31, 49, 58, 61, 64, 70, 77, 88, 97–99, 101, 105, 109, 121, 127, 155, 156, 157, 158
  chondrule abundance 157
  chondrule diameter 14, 16
  $\delta^{17}O$ 58, 59, 101
  $\delta^{18}O$ 58, 59, 99, 101
  Fe/Si 7
  Fe_m/Fe_t 1, 14
  magnetic 8, 31, 61, 111, 114, 116, 117, 145
  matrix abundance 58, 152
  metal abundance 88
  metal grain size 58, 138
  Mg/Si 127, 128, 129
  petrographic types 19, 39, 63, 80, 97
  physical 2, 9, 21, 31, 39, 43, 51, 59, 60, 65, 69, 94, 101, 107, 126, 135, 152, 157, 159
  shock-blackened 32, 75
  water 12, 15, 19, 21, 22, 31, 40, 57, 60, 64, 94, 97, 104, 119, 121, 122, 139, 148, 151, 152, 155–157
chondrite texture 60, 77, 121
  chondrule sizes 58, 70, 133
chondrites 6, 9, 10, 11, 12, 15, 19, 20, 21, 25, 29, 32, 34, 35, 39, 49, 50, 51, 53, 54, 55, 57–61, 63, 66, 67, 70, 71, 73, 77, 80, 85, 93, 94, 96, 97, 98, 101, 105, 106, 107, 109, 111, 119, 121, 122, 126, 127, 130–134, 137, 140, 141, 145, 149, 151, 152, 154–159
  links between groups 21
  links with iron meteorites 1, 27, 35, 50, 70, 166, 172, 179, 190
  metal size 14
  metal-rich 49, 77, 123, 146, 158
chondrule abundance 7, 157
chondrule ages 104
  I–Xe ages 68, 94
chondrule classes 20, 49
  barred olivine 77
  cryptocrystalline 77
  Dodd scheme 77
  droplet 85, 105, 112, 114, 117, 122, 123, 129, 130, 133, 138, 139, 152
  Gooding–Keil scheme 77
  group A1 82, 88, 89, 100, 183, 191
  group A2 105
  group A3 176
  group A5 80, 82, 83
  group B1 82
  groups 79, 80
  Kieffer–King scheme 75
  lithic 61, 75, 77
  Merrill scheme 75
  non-porphyritic 75, 77
  porphyritic olivine–pyroxene 25, 77
  porphyritic pyroxene 77
  radiating pyroxene 75
  Scott–Taylor–Jones scheme 39, 77, 79, 105
  Sears *et al.* scheme 79, 75
  Tschermak scheme 75
  type I 17, 79
  type II 79
  type IIA 79, 93
chondrule composition 79, 82, 84, 86, 88, 92, 96, 135, 151
  Al 79, 80, 82, 83, 85, 88, 89, 93, 105, 183
  boron isotopes 102
  Ca 51, 57, 61, 85, 92, 93, 118
  complementary composition of components 134
  Cr isotopes 82, 105
  factor analysis 82, 84
  Fe 2, 3
  Fe isotopes 51, 53, 88, 145
  FeO 18
  H isotopes 53, 55, 98
  isotopes 8, 58, 63, 64, 67
  K 2
  K isotopes
  laboratory experiments 59, 69, 82, 83, 85, 90, 92, 100, 110, 116
  laboratory heating 86
  Mg 14, 18, 39, 51, 57, 61, 67, 85, 86, 92, 117, 118, 122, 127
  Mg isotopes 51, 85
  Na 1, 3, 4
  open vs closed system 58, 88
  oxygen fugacity 126, 134
  Si 2–12
  Si isotopes 57, 85, 127
  Ti 57
  Xe isotopes 103
chondrule diameter 14, 16
chondrule distribution 3, 5, 6, 10, 12, 15, 16, 17, 20, 35–37, 47, 50, 58, 54, 55, 57, 58, 60, 63–65, 70, 71, 73–75, 90, 91, 92
chondrule history 84, 90
  aqueous alteration 15, 60, 61, 63–65, 88, 93, 94, 103
  cooling rates 110, 114, 130, 133, 135, 158
  crystallization 36, 37, 60, 85, 87, 90, 92, 110, 121
  density sorting 151, 156
  metamorphism 15, 17, 19, 35, 54, 55, 60, 61, 63, 66, 79, 80, 93, 128, 137, 156, 157, 158
  nucleation centers 90, 92
  recycling 88, 93, 152
  sorting 16, 53, 61, 118, 123, 133, 138, 139, 141–144, 146, 147, 151, 156–158
  supercooling 53, 85
  temperatures 36, 37, 39, 53, 57, 61, 63, 87, 90, 92, 93, 114, 127, 133, 144, 145, 157
chondrule mesostasis 94, 97
chondrule metamorphism 93
chondrule olivine 36, 75, 79
chondrule origin 122
  droplets of fiery rain 6, 112, 121
  early studies 60
  formation 10, 13, 20, 31, 47, 55, 57, 58, 66, 67, 70, 82, 85, 94, 98, 102, 104, 106, 111, 112, 114, 117, 119
  impact 10, 11, 15, 20, 22, 25, 29, 31, 35, 37, 38, 39, 40, 41, 43, 44, 45, 46, 47, 60, 61, 88, 109–114, 119–123, 129, 130, 133–135, 137–140, 143, 148, 151–159

chondrule origin (*cont.*)
    non-asteroidal
    null arguments against impact origin 130, 137
    parent bodies 10, 20, 21, 30, 32, 34, 35, 39, 70, 95, 119, 123, 132, 138, 151
    primordial solar nebula 55, 61, 97, 106, 112, 113, 116–118, 125, 127, 143, 144, 145
chondrule processes 119, 143
    open systems 88, 89, 151
chondrule properties 70, 89, 90
    classes 6, 7, 8, 11, 14, 15, 16, 18, 20, 25, 27, 29, 30, 34, 49, 50, 54, 55, 58, 59, 63, 70, 73, 77, 80, 82, 83, 85, 93, 98, 105, 106, 107, 109, 115, 127, 129, 132, 133, 134, 139, 141, 146, 151, 153, 152, 158
    distribution over classes 22
    diversity 73, 121, 131, 133, 151, 156
    EH 14, 16, 49, 50, 53, 54, 63, 70, 89, 105, 156, 158
    interclass comparisons 105
    links with refractory inclusions
    lunar 3, 4, 10, 11, 31, 32, 90, 109, 117, 122, 130, 136, 139, 140, 143, 159
    number density 133
    properties 3, 6, 8, 11, 14, 15, 21, 27, 29, 30, 32, 34, 35, 43, 45, 49, 50, 51, 57, 58, 60, 61, 64, 69, 70, 73, 75, 79, 82, 85, 86, 88, 89, 90, 93, 95, 101, 102, 106, 107, 112, 114, 122, 126, 127, 128, 131, 133, 134, 141, 152, 156, 159
    rims 73, 75, 88, 89, 94, 95, 96, 97, 98, 101, 107, 110, 114, 130
    sizes 14, 15, 39, 50, 58, 70, 133, 141, 144, 157
chondrule sorting 141
    abrasive processes 143
    aerodynamic processes 142
    ballistic processes 142
chondrule texture 73, 75, 77, 90, 92
    armoured 94
    barred 77, 93, 104, 109, 123
    barred olivine 77, 93, 104, 109, 123
    clastic 77, 82, 123, 151, 153, 154, 155, 156, 158
    compound 21, 75, 77, 116, 130, 133
    cryptocrystalline 77
    droplet 6, 31, 73, 75, 82, 85, 105, 112, 114, 117, 122, 123, 129, 130, 133, 138, 139, 152
    exocentroradial
    fayalite borders 94
    glassy 31, 64, 77, 110
    lithic 61, 75, 77, 82, 105, 143
    metal-poor microporphyritic 77, 82, 123
    metal-rich microporphyritic 82, 123
    non-porphyritic 75, 77, 82, 93, 121
    plagioclase-rich 75, 107
    porphyritic 75, 77, 78, 82, 93, 104, 121, 123
    porphyritic olivine 75, 77, 104
    porphyritic olivine–pyroxene 77
    porphyritic pyroxene 77, 82, 104
    radiating 25, 75, 77, 104, 109
    radiating pyroxene 75, 77, 104, 109
    relic grains 92, 93
chronology 8
    K–Ar 34, 67
    Rb–Sr 9, 10

CI chondrites 7, 8, 49, 54, 59, 63, 83, 85, 119, 127, 131, 133, 140, 147, 155, 156, 157, 158
    aqueous alteration 15, 60, 61, 63, 64, 65, 88, 93, 94, 103
    components 1, 3, 15, 16, 19, 50, 55, 58, 60, 61, 66, 84, 127, 129, 133, 134, 135, 151, 152, 154, 156, 157
    densities 21, 22, 23, 30, 40, 43, 114, 116, 119, 127, 133, 134, 139, 143, 148
    formation 10, 11, 13, 16, 20, 31, 32, 34, 35, 40, 43, 45, 47, 50, 55, 57, 58, 60, 66, 67, 68, 64, 70, 73, 75, 77, 82–90, 93–98, 100–106, 111–160
    matrix 1, 3, 5, 14, 15, 16, 18, 27, 30, 31, 49, 55, 58, 60, 64, 75, 87, 94, 96, 97, 98, 134, 135, 151, 152, 156
    metal–silicate fractionation 7, 20, 49, 54, 55, 70, 71, 141
    O isotopes 90
    petrographic type 3, 17, 18, 19, 36, 37, 38, 39, 54, 55, 63, 80, 93, 97
    water 12, 14, 15, 19, 21, 22, 23, 31, 40, 57, 59, 60, 64, 94, 97, 104, 119, 121, 122, 139, 147, 148, 151, 152, 155, 156, 157
CM chondrites 8, 16, 19, 23, 49, 61, 63, 64, 96, 97, 106, 109, 131, 133, 140, 156, 157
    aqueous alteration 15, 60, 61, 63, 64, 65, 88, 93, 94, 103
    components 1, 3, 15, 16, 19, 50, 55, 58, 60, 61, 66, 84, 127, 129, 133, 134, 135, 151, 152, 154, 156, 157
    densities 21, 22, 23, 30, 40, 43, 114, 116, 119, 127, 133, 134, 139, 143, 148
    formation 10, 13, 16, 20, 31, 34, 47, 55, 57, 58, 60, 61, 63, 65–70, 73, 82–90, 93–95, 98, 100, 102–106, 111–148, 151–154, 157, 159
    matrix 1, 3, 5, 14, 15, 16, 18, 27, 30, 31, 49, 55, 58, 60, 64, 75, 87, 94–98, 134, 135, 151, 152, 156
    metal–silicate fractionation 7, 20, 49, 54, 55, 70, 71, 141, 143, 145, 147, 149, 152, 156–158
    O isotopes
    petrographic type 17, 18, 36, 37, 38, 39, 54, 55, 63, 93
    water 12, 14, 15, 19, 21, 22, 23, 31, 40, 57, 59, 64, 94, 97, 104, 119, 139, 147, 148, 151, 152, 155, 156, 157
CO chondrites 16, 19, 49, 63, 81, 98, 99, 157
    components 1, 3, 15, 16, 19, 50, 55, 58, 60, 61, 66, 84, 127, 129, 133, 134, 135, 151, 152, 154, 156, 157
    formation 10, 13, 16, 20, 31, 34, 47, 55–70, 73, 82–90, 93–95, 98, 100, 102–106, 111–114, 117, 119, 120–152, 154, 157, 159
    matrix 1, 3, 5, 14, 15, 16, 18, 27, 30, 31, 49, 55, 58, 60, 64, 75, 87, 94, 95, 96, 97, 98, 134, 135, 151, 152, 156
    metal–silicate fractionation 7, 20, 49, 54, 55, 70, 71, 141, 143, 145, 147, 149, 152, 156, 157, 158
    petrographic type 17, 18, 36, 37, 38, 39, 54, 55, 63, 93, 97
    thermal history 35, 37, 63

comets 4, 11, 13, 152, 154, 155
  formation 10, 13, 16, 20, 31, 34, 47, 55, 57, 58, 60, 61, 63, 65, 66, 67, 68, 69, 70, 73, 82, 83, 84, 85, 86, 87, 88, 89, 90, 93, 94, 95, 98, 100, 102, 103, 104, 105, 106, 111, 112, 113, 114, 117, 119–157, 159
comminution 44
condensation sequence 57, 129
cooling rates 17, 35, 37, 62, 63, 90, 92, 93, 110, 114, 130, 133, 135, 158
chondrules
cosmic ray exposure ages 32, 33, 69, 70
cosmic rays 32, 33
CR chondrites 49, 107
Craig, Harmon 6
crater densities 43
craters 11, 41, 43, 44, 135–138, 140, 160
  asteroid 4, 10, 11, 12, 13, 21–45, 47, 66, 69, 112–127, 140, 142, 145, 148, 151–161
  Earth 3, 4, 9, 10, 11, 12, 21, 22, 24, 25, 28, 29, 32, 43, 65, 69, 107, 114, 115, 116, 118, 127, 128, 130, 131, 132, 145, 155, 158, 160, 161, 162
  Mars 4, 12, 40, 42, 43, 125
  Moon 3, 4, 10, 12, 20, 26, 30, 31, 35, 40, 45, 69, 106, 109, 122, 130, 135, 140, 145, 156, 159, 162
craters, named
  Himeros 41, 43
  Psyche 22, 41, 43
  Ries 1, 3, 4, 6, 12, 15, 17, 19, 35, 39, 45, 49, 64, 71, 75, 79, 80, 85, 87, 88, 95, 105, 110, 111, 112, 115, 119, 123, 129, 130, 131, 137, 141, 143, 158, 159, 160, 161
  Shoemaker 11, 40, 41, 42, 43
crystallized lunar 109, 143
  spherules (CLS) 109
Curie point 61, 145
CV chondrites 16, 49, 63, 106, 107, 157
  aqueous alteration 15, 60, 61, 63, 64, 65, 88, 93, 94, 103
  chondrules 1, 3, 6, 10, 12, 15, 20, 35, 36, 47, 50, 53, 55, 58
  components 1, 19, 50, 55, 58, 60, 61, 66, 127, 129, 133, 134, 135, 151, 152, 157
  formation 10, 13, 16, 20, 31, 34, 47, 55–73, 82–106, 111–114, 117, 119–139, 141, 143, 147, 148, 151, 152, 154, 157, 159
  inclusions 1, 3, 16, 59, 65–69, 86, 101, 102, 105, 106, 107, 109, 120, 128, 129, 156, 157
  metal–silicate fractionation 7, 20, 49, 54, 55, 70, 71, 141, 143, 145, 147, 149, 152, 156–158
  O isotopes
  petrographic type 17, 18, 36, 37, 38, 39, 54, 55, 63, 93, 97
  refractory inclusions 1, 3, 16, 59, 65, 66, 67, 69, 86, 101, 102, 105, 106, 107, 128, 129, 156, 157

$\delta^{17}O$ 14, 59, 99, 101, 102
$\delta^{18}O$ 14, 59, 99, 101, 102
Dactyl 42
dark inclusions 109
degassing 32, 151

density of chondrites 131
Diemos 40, 42
drag 118, 119, 133, 142, 146, 148, 151
  aerodynamic 118, 119, 133, 139, 141–147, 151, 156
  Epstein 142, 146
  Stokes 142, 146
droplets 6, 31, 73, 112, 114, 122, 123, 129, 130, 133, 138, 139, 152
  ablation 103, 111, 114
dust mantles 98

E chondrites 6, 11, 12, 15, 19, 20, 27, 29, 37, 49, 50, 58, 59, 66, 67, 81, 97, 105, 106, 122, 126, 127, 130, 137, 141, 156, 157, 158
  I–Xe ages 94
Earth 3, 4, 9, 10, 11, 12, 21, 22, 24, 25, 28, 29, 32, 43, 65, 69, 107, 114–116, 118, 127, 128, 130–132, 145, 155, 158, 160–162
EH chondrites 53, 63, 105, 158
  components 1, 3, 15, 16, 19, 50, 55, 58, 60, 61, 66, 84, 127, 129, 133, 134, 135, 151, 152, 154, 156, 157
  metal–silicate fractionation 7, 20, 49, 54, 55, 70, 71, 141, 143, 145, 147, 149, 152, 156, 157, 158
  thermal history 35, 37, 63
ejecta 31, 36, 43, 44, 45, 47, 109, 113, 122, 138, 139, 140, 160, 161
  fraction escaping 44
  properties 3, 6, 8, 11, 15, 21, 22, 29, 30, 32, 34, 35, 43, 45, 49, 50, 51, 52, 54, 57, 58, 60, 61, 64, 69, 70, 73, 75, 79, 82, 85, 86, 89, 93, 95, 101, 102, 107, 112, 114, 122, 126–128, 131, 133, 134, 141, 155, 156, 159
  velocity 47, 115, 118, 119, 122, 123, 130, 138, 142, 143, 149
EL chondrites 49, 53, 63, 70, 158
  components 1, 19, 50, 55, 58, 60, 61, 66, 127, 129, 133, 134, 135, 151, 152, 157
  metal–silicate fractionation 7, 20, 49, 54, 55, 70, 71, 141, 145
  thermal history 35, 63
element ratios 15
  C/Si 158
  Ca/Si 106, 127
  Fe/Si 7
  Mg/Si 127, 128, 129
elements 6, 8, 12, 13, 15, 24, 35, 50, 51, 53, 54, 55, 57, 58, 82, 85, 88, 95, 103, 106, 107, 117, 126, 127, 128, 129, 134, 146
  Al 39, 57, 62, 63, 67, 69, 85, 94, 97, 98, 101, 104, 106, 107, 118, 123, 137, 151
  Ar 32, 34, 51, 67, 70, 132
  atmophile 51
  Bi 12
  Ca 51, 57, 61, 79, 85, 92, 93, 105, 106, 118, 127
  chalcophile 7, 51, 82, 145
  Co 51, 145
  Cr 82
  Cu 51
  Fe 6, 7, 51, 53, 55, 57, 61, 77, 79, 82, 83, 85, 86, 88, 89, 93, 94, 101, 102, 105, 110, 115, 118, 122, 145

elements (*cont.*)
  Ga 9, 32, 43, 44, 65, 67, 82, 85, 145
  highly volatile 51, 55, 85
  In 51, 55, 69
  Ir 51, 55, 145
  K 34, 39, 49, 57, 63, 64, 66, 67, 82, 88, 92, 101, 102, 115, 127, 144, 145, 155
  lithophile 51, 55, 82, 134, 145, 152, 158
  major 4, 6, 9, 10, 15, 21, 24, 25, 30, 32, 34, 35, 43, 50, 57, 59, 64, 67, 73, 75, 79, 80, 82, 85, 89, 90, 99, 106, 132, 135, 138, 139, 152, 157, 159, 162
  Mg 39, 51, 57, 61, 67, 85, 92, 118, 122, 127, 128, 129
  Mn 67, 82, 105, 110
  Mo 107
  moderately volatile 51, 55, 57, 129
  N 4, 6, 51, 57, 88, 89, 145
  Na 51, 57, 82, 85, 88, 89, 101, 103
  Ne 51
  Ni 17, 51, 63, 145
  refractory 31, 51, 53, 57, 65–67, 69, 82, 83, 85, 88, 92, 101, 105–107, 127–129, 134, 157
  Sb 82, 85
  Se 82, 85
  Si 51, 57, 85, 106, 118, 127, 128, 129, 135, 137, 158
  siderophile 51, 53, 54, 55, 145
  Tl 51, 55, 157
  volatile 31, 35, 45, 51, 53, 54, 55, 57, 58, 82, 85, 88, 95, 96, 98, 103, 107, 119, 122, 123, 129, 139, 147, 152, 153, 157, 159
  W 107
  Xe 51, 67, 94, 103, 104, 160
  Zn 82, 85
enstatite chondrites 20
Epstein drag *see* drag, Epstein
Ergun equation 148, 149
Eros 21, 22, 40, 41
  boulders 41
  crater 41
  grooves 41, 43
  regolith 11, 19, 30, 31, 32, 35, 36, 39, 41
  ridge 41
evaporation sequence 57, 107, 129
exposure ages 32, 33, 69, 70, 132
  H chondrites 9, 24, 25, 29, 32, 33, 39, 70, 132, 145
  L chondrites 6, 25, 32, 33, 34, 70, 122, 135, 145
  LL chondrites 32, 33, 49, 53, 55, 59, 98, 132

factor analysis 82, 84
Fe/Si 7, 8
Fe_m/Fe_t 14
fluidization 148, 151
formation interval 67
FUN inclusions 101, 157

gehlenite 57, 106
grains 1, 12, 17, 18, 31, 32, 35, 37, 39, 44, 55, 60, 69, 70, 92, 93, 94, 95, 96, 97, 100, 107, 109, 113, 117–119, 123, 126, 127, 130, 133–135, 137, 141–147, 152, 154, 157

isolated olivine 96
presolar 69, 154
gravitational energy 131

H chondrites 9, 24, 25, 29, 32, 33
  ages 32, 33, 34, 39, 61, 63, 65, 67–69, 70, 94, 104, 130, 132, 137
  components 1, 3, 15, 16, 19, 55, 58, 59–61, 66, 84, 127, 129, 133, 134, 135, 151, 152, 154, 156, 157
  cooling rates 17, 35, 37, 62, 63, 90, 92, 93, 110, 114, 130, 133, 135, 158
  cosmic ray exposure 32, 33, 69, 70
  definition 79
  exposure ages 32, 33, 69, 70, 132
  I–Xe ages 94
  metal–silicate fractionation 7, 20, 49, 54, 55, 70, 71, 141, 145
heat source 37, 38, 39, 67, 129
holy smoke 97
Howard, Edward Charles 1
howardite 26, 27, 109, 110
Humboldt, Alexander von 19, 94, 97

impact 10, 11, 15, 20, 25, 31, 35, 37–47, 60, 61, 88, 109–113, 117, 119, 120, 122, 123, 129, 130, 133–135, 137–140, 143, 148, 151–153, 155–159
  heating 20, 27, 32, 46, 47, 57, 61, 69, 73, 85, 87, 92, 109, 110, 116, 117, 118, 119, 120, 121, 122, 123, 129, 130, 138, 139, 152, 156
  melt 27, 32, 46, 47, 57, 61, 69, 73, 85, 87, 92, 109, 110, 116, 117, 118, 119, 120, 121, 122, 123, 129, 130, 138, 139, 152, 156
  melt production 46, 97, 122
  melt rocks 20
  melt spherules 47, 109, 110, 120, 122, 130, 152
  origin of chondrules 10, 71, 111, 122, 139, 156
  pyroclastics 151, 153, 158
inclusions 3, 16, 59, 65, 66, 67, 69, 86, 101, 102, 105, 106, 107, 109, 120, 128, 129, 156, 157
  refractory 3, 16, 31, 51, 53, 54, 57, 59, 65, 66, 67, 69, 82, 83, 84, 85, 86, 87, 88, 92, 95, 101, 102, 105, 106, 107, 127, 128, 129, 134, 156, 157
instruments 4, 8, 10, 73, 111
  activation analysis 8, 50, 80, 81
  defocussed electron microprobe analysis 81
  INAA 50, 80, 81
  instrumental analysis 4
  isotope dilution mass spectrometry 82
  mass spectroscopy 8
  optical microscopy 4
  reflected light microscopy 4, 6
  transmitted light microscopy 4
  wet chemical analysis 50
  X-ray diffraction 7
  X-ray fluorescence 7, 50
interstellar dust 113, 118, 154
iodine-129
  chondrules
iron meteorites 1, 27, 29, 35, 50, 70
  I–Xe ages 94
iron sulfide 57

ISAS 160, 161
isotopes 8, 58, 63, 64, 67, 101, 102, 103, 106, 153, 158
^{26}Al 38, 39, 67, 69, 104, 106, 123, 151
^{39}Ar 66
^{11}B 102
^{56}Fe 102
^{129}I 67, 68, 104
^{40}K 64, 66
^{41}K 102
^{26}Mg 39, 67
^{53}Mn 67, 105
O 90
^{16}O 58, 59, 99, 100, 101, 157
^{17}O 58, 59
^{18}O 58, 59
^{244}Pu 61, 63, 67
^{87}Rb 9, 66
^{87}Sr 9, 10, 64, 66
^{129}Xe 67, 68, 104

Jupiter 10, 24, 25, 31, 42, 117, 123, 125, 127, 137, 138

kamacite 18, 35, 61
Kirkwood Gaps 24, 25

L chondrites 6, 25, 32, 33, 34, 70, 122, 132, 145
   age 1, 9, 10, 31, 32, 33, 63–66, 68, 69, 104, 130
   components 1, 3, 15, 16, 19, 55, 58–61, 66, 84, 127, 129, 133, 134, 135, 151, 152, 154, 156, 157
   cosmic ray exposure 32, 33, 69, 70
   definition 79
   exposure ages 32, 33, 69, 70, 132
   I–Xe ages 94
   K–Ar ages 34, 68
   metal–silicate fractionation 7, 20, 49, 54, 55, 70, 71, 141, 145
   O isotopes
Lavoisier, Antoine 1, 2
lightning 1, 2, 111–116
   origin of meteorites 3, 122
LL chondrites 27, 32, 33, 49, 53, 55, 59, 98, 132
   components 1, 3, 15, 16, 19, 55, 58, 59, 60, 61, 66, 84, 127, 129, 133, 134, 135, 151, 152, 154, 156, 157
   cosmic ray exposure 32, 33, 69, 70
   definition 79
   exposure ages 32, 33, 69, 70, 132
   I–Xe ages 94
   metal–silicate fractionation 7, 20, 49, 54, 55, 70, 71, 141, 145
lunar chondrules 109, 117, 140
lunar meteorites 109
magnetics 8, 31, 61, 111, 114, 116, 117, 145
Main Belt asteroids 21, 25
   distribution over classes 22
manganese-26
matrix 1, 3, 5, 14–16, 18, 27, 30, 31, 49, 55, 58, 60, 64, 75, 87, 94, 95, 96, 97, 98, 134, 135, 151, 152, 156
   abundance 7, 8, 9, 14, 15, 27, 33, 53, 54, 55, 57, 58, 60, 68, 73, 80, 82, 83, 88, 89, 90, 104, 109, 128, 129, 140, 145, 152, 156, 157

Merapi Volcano 154
mesostasis 64, 75, 79, 80, 87, 88, 89, 92, 93, 94, 97, 100, 101
   composition 6, 7, 11, 13, 15, 22, 31, 37, 39, 44, 49–51, 53, 54, 56–58, 60, 73, 77, 79, 80, 83, 88, 92, 96, 97, 99, 101, 107, 120, 125, 126, 127, 130, 133, 134, 135, 140, 147, 152, 156
   metal 3, 5, 6, 7, 8, 14, 15, 16, 17, 20, 27, 29, 32, 35, 37, 39, 43, 50–51, 54–58, 61, 70, 71, 73, 77, 82, 84, 88, 94, 95, 97, 107, 121, 123, 127, 133–135, 138, 141, 143–153, 155–159
   abundances 7, 8, 14, 15, 39, 51, 53, 54, 55, 58, 127, 145, 148
metal grain size 14, 16, 138
metal–silicate fractionation 7, 20, 49, 54, 55, 70, 71, 141, 145
   aerodynamics 142, 144, 145
   condensation and settling 144, 146
   crystal growth 143, 144
   ductility 144, 145
   magnetism 61, 144, 145
   nebula 35, 50, 53, 55, 57, 59, 60, 61, 64, 65, 88, 95, 97, 98, 106, 107, 111–119, 125–127, 129–131, 133–135, 137, 138, 142–147, 151, 159, 160
   parent body 11, 25, 29, 31, 32, 33, 35, 36, 38, 39, 50, 61, 63, 64, 65, 67, 69, 94, 104, 112, 120–122, 127, 129, 134, 138–140, 143, 146, 147, 156, 158
metamorphism 15, 17, 19, 35, 54, 55, 60, 61, 63, 66, 79, 80, 93, 129, 137, 156, 157, 158
   age 1, 9, 10, 31, 32, 33, 63, 64, 65, 66, 68, 69, 104, 130
   cooling rates 17, 35, 37, 62, 63, 90, 92, 93, 110, 114, 130, 133, 135, 158
   olivine Ca 92
   olivine Fe 94
   petrologic types 1, 9, 35, 36, 37
   temperatures 35, 36, 37, 39, 53, 57, 61, 63, 87, 90, 92, 93, 114, 127, 133, 144, 145, 157
meteorite classes 7, 50
   asteroid links 21, 29
meteorites 1–6, 8–13, 15, 17, 19–22, 25–29, 31, 32, 34, 35, 37, 39, 45, 47, 49, 50, 51, 57–62, 64–70, 73, 79, 82, 83, 98, 101, 105, 107, 109, 110, 116, 117, 120–122, 127, 131, 132, 134, 143, 146, 148, 151, 152, 154–161
   asteroid links 21, 29
   atmospheric passage 131, 132
   distribution 5, 22, 28, 29, 39, 51, 58, 63, 92, 104, 105, 109, 116, 126, 141, 151, 156
   early studies 60
   expeditions 11, 12
   formation 10, 13, 16, 20, 31, 34, 47, 55, 57, 58, 60, 66–70, 73, 82–90, 93–95, 98, 100, 102–106, 111–114, 117, 119–131, 134–143, 147, 148, 151, 152, 154, 157, 159
   lunar 3, 4, 10, 11, 31, 32, 90, 109, 117, 122, 130, 136, 139, 140, 143, 159
Acfer 059 146
Alais 23
ALH 81302 23
ALH 83100 23

meteorites (*cont.*)
　ALHA76008 62
　ALHA77262 62
　ALHA77294 62
　ALHA78076 62
　ALHA78111 62
　ALHA78115 62
　ALHA78134 62
　ALHA79026 62
　ALHA79035 62
　ALHA80121 62
　ALHA80131 62
　ALHA81092 62
　ALHA81105 62
　Allan Hills 84001 12
　Allan Hills 85085 49
　Allegan 62, 63
　Allende 66, 81, 96, 101, 102, 103, 106, 107
　Ankober 62
　Barratta 77
　Barwell 17
　Beaver Creek 62, 77
　Bells 64
　Benten 23
　Bhola 62
　Bishunpur 17, 97
　Chainpur 62, 81
　Chateau-Renard 75
　Cold Bokkeveld 23, 64
　Conquista 62
　Cullison 77
　Cynthiana 17
　Dhurmsala 75
　Elm Creek 77
　Erakot 23
　Essebi 23
　Estacado 62
　Fayetteville 30
　Forest Vale 62
　Guarena 62, 63, 66
　H chondrites 9, 24, 25, 29, 32, 33, 39, 70, 132, 145
　Haripura 23
　Hendersonville 77
　Hessle 77
　Homestead 9, 75
　Ivuna 23
　Kakangari 50
　Kapoeta 109
　Kernouve 62, 63
　Kesen 62
　Knyahinya 9, 75
　Krymka 62, 79, 81
　Mezo Madaras 19, 75
　Mighei 23, 64
　Monroe 75
　Mt. Browne 62
　Murchison 23, 64, 65, 96
　Murray 5, 23, 64, 81
　Nadiabondi 62
　Nawapali 23
　Netschaevo 109
　Nogoya 64
　Nuevo Mercurio 62
　Olivenza 62
　Orgueil 23
　origin 2, 3, 10, 15, 20, 49, 60, 70, 71, 98, 109, 111, 114, 121, 122, 130, 135, 140, 141, 156, 159, 160
　parent bodies 10, 20, 21, 30, 32, 34, 35, 39, 40, 70, 94, 119, 123, 132, 138, 151
　Parnallee 77, 81
　Pavlovka 109
　Peace River 17, 66
　Pollen 64
　Ramsdorf 122
　Renazzo 53, 114, 115
　Revelstoke 22
　Richardton 62, 63, 81
　Roosevelt County 5
　Santa Cruz 23
　Semarkona 5, 64, 78, 79, 81, 82, 83, 84, 94
　Sena 62
　Shelburne 17
　Sienna
　St. Mesmin
　St. Séverin 62, 65
　Staroje Boriskino 23
　Ste. Marguerite 63
　Tagish Lake 22
　Tennasilm 77
　Tieschitz 94, 97
　Tuxtuac 62
　Uden 62
　Mg/Si 14, 127, 128, 129
　　Yamato 791824 23
　microchondrules 94
　　Yamato 793321 23
　microcraters 135, 136, 137
　minerals 4, 5, 6, 7, 18, 30, 31, 34, 44, 49, 57, 58, 64, 67, 94, 97, 99, 102, 103, 106, 127
　　corundum 57
　　enstatite 16, 20, 27, 49, 53, 56, 57, 58, 59, 81, 105, 106, 114, 118, 123, 127, 157, 158
　　fayalite 18, 56, 57, 94
　　ferrosilite 56, 57
　　forsterite 53, 56, 57, 90, 114
　　gehlenite 57, 106
　　hibonite 57, 106
　　olivine 18, 25, 27, 32, 36, 53, 64, 65, 75, 77, 79, 80, 82, 84–88, 90, 92–97, 99, 100, 104–107, 109, 114, 115, 121, 122, 136, 155
　　orthopyroxene 27, 57, 61
　　pyroxene 18, 25, 27, 57, 61, 75, 77, 78, 79, 82, 84, 85, 86, 88, 90, 92, 99, 104, 105, 106, 107, 109, 121
　Moon 3, 4, 10, 12, 20, 26, 30, 31, 35, 40, 41, 42, 43, 44, 45, 69, 106, 109, 122, 130, 135, 140, 145, 156, 159, 162
　　Apollo program 10, 30
　　pyrrhotite 56, 57
　　spinel 57, 97, 106, 107
　Mount Saint Helens 152
　　lunar samples 10, 11, 109, 159

NASA 4, 12, 41, 136, 162
National Aeronautics and Space Administration *see* NASA
NEAR Shoemaker spacecraft 11, 41, 43
near-Earth asteroids 11, 21, 161, 162
nebula 35, 50, 53, 55, 57, 58, 59, 60, 61, 64, 65, 88, 95, 97, 98, 106, 107, 111–119, 125–138, 142–147, 151, 159, 160
  chemical processes 44, 127
  distribution 5, 22, 28, 29, 32, 39, 51, 58, 63, 92, 104, 109, 116, 126, 141, 151, 156
  pressure 32, 46, 47, 51, 53, 55, 67, 85, 87, 114, 115, 117, 119, 125, 126, 127, 128, 144
nebular theories for chondrule formation 129, 130
  aerodynamic drag heating 118
  critique 129, 131, 133, 135
  direct condensation 113, 114, 122
  geographical factors 129
  impact within Jovian protoplanet 117
  impacts between grains 117
  interstellar processes 118
  lightning 1, 2, 111, 112, 113, 115, 116
  magnetic field 61, 113, 116, 117
  primordial solar 55, 61, 97, 106, 112, 113, 115–118, 125, 127, 143–145
nickel 1, 6, 17, 18
  shock waves 113, 117, 119
nucleation centers 90, 92
  solar activity 112
nuclides 32, 33, 39, 64, 67, 69
nuclides, long-lived
  cosmogenic 32, 69
  ^{40}K 400
  ^{147}Sm 64
nuclides, short-lived 67
  ^{26}Al 39, 67, 69, 104, 123, 151
  ^{129}I 67, 104
  ^{53}Mn 67, 105
  ^{107}Pd 67
  ^{244}Pu 61, 63, 67
  ^{87}Rb 9
  ^{146}Sm 67
  U 39, 64
nuée ardentes 140

OCL line 157
olivine 18, 25, 27, 32, 36, 53, 64, 65, 75, 77, 79–82, 84–88, 90, 92–94, 96, 97, 99, 100
  Ca 14, 18, 51, 54, 57, 61, 85, 92, 93, 105, 106, 118, 127
  CaO 79
  Fe 6, 7, 8, 14, 27, 51, 52–57, 61, 77, 79, 83–88, 93, 94, 105, 110, 121, 122, 145, 146, 152
  FeO 18, 51, 53, 78, 82, 83, 85, 87, 88, 89, 92, 93, 97, 101, 115, 118, 122
Olmsted, H. 3
onion skin structure 36, 37, 38, 39, 63
orbits 24, 25, 31, 154
ordinary chondrites 16, 17, 19, 21, 26, 27, 28, 29, 33, 49, 53, 55, 58, 59, 61, 63, 70, 77, 79, 80, 81, 86, 94, 97, 98, 99, 101, 103, 105, 107, 131, 132, 146, 149, 157, 155, 157
  atmospheric passage 131, 132
  matrix 1, 3, 5, 14, 15, 16, 18, 27, 30, 31, 49, 55, 58, 60, 64, 75, 87, 94–98, 134, 135, 151, 152, 156
O isotopes
  petrographic type 17, 18, 36, 37, 38, 39, 54, 55, 63, 93, 97
oxygen isotopes 63, 101, 153, 158, 162
  Allende inclusions 66, 101
  chondrules 3, 5, 6, 10, 12, 15–17, 20, 35, 36, 37, 47, 50, 53, 55, 57, 58, 60, 63, 64, 70, 71, 73–162
  ordinary chondrite chondrules 101, 162

parent bodies 10, 20–40, 42, 44, 46, 70, 95, 119, 121, 123, 132, 138, 151, 162
  cooling rates 17, 35, 37, 62, 63, 90, 92, 93, 110, 114, 130, 133, 135, 158, 162
  onion skin structure 36, 37, 38, 39, 63, 162
  rubble pile structure 37, 39, 63, 162
  structure 7, 11, 17, 31, 35, 36, 37, 38, 39, 43, 50, 63, 92, 121, 122, 139, 162
parent body theories for chondrule formation 11, 31
  impact 10, 11, 15, 20, 22, 25, 29, 31, 35, 37–47, 60, 61, 88, 109, 110, 111–113, 117, 119, 120, 122, 123, 129, 130, 133–140, 143, 148, 151–159, 162
  volcanism 66, 112, 119, 120, 122, 152, 153, 156, 162
Pele's Hair 112
petrographic types (petrologic types) 39, 55
  origin of meteorites 3, 122
phase diagrams 53, 85, 87
Phobos 40, 42
pits 130, 137
  impact 10–162
planetesimals 13, 57, 60, 123, 125
planets 13, 112, 121, 125, 126, 127, 154
  formation 10, 11, 13, 16, 20, 31, 32, 34, 35, 40, 43, 45, 47, 50, 55–73, 75, 77, 82–90, 93, 94, 95, 98, 100, 102–106, 111–131, 134–139, 141–160
porosity 21, 22, 23, 35, 46, 139, 152
  chondrites 6–86, 93–111, 119, 121, 122, 126, 127, 130–134, 140, 141, 145–149, 151–162
precursor material 54, 87, 154, 155
Prior, George Thurland 6
Prior's Law 6
protoplanets 125
proto-Sun 125

R chondrites 49, 107
rare earth elements 107, 128
  refractory inclusions 3, 16, 59, 65–69, 86, 101, 102, 105–107, 128, 129, 156, 157
Rayleigh equation 102
redox (reduction–oxidation) 152
reduction–oxidation series 6
refractory inclusions 3, 16, 59, 65, 66, 67, 69, 86, 101, 102, 105, 106, 107, 128, 129, 156, 157
  amoeboid–olivine aggregates 106
  anorthite-rich inclusions 1, 106

refractory inclusions (*cont.*)
  CI 7, 8, 14, 16, 19, 22, 23, 27, 40, 49, 54, 58, 59, 60, 63, 82, 83, 85, 97, 105, 119, 121, 127, 131, 133, 140, 147, 148, 151, 152, 155, 156, 157, 158
  CG14 101, 102
  CM chondrites 8, 16, 19, 23, 49, 61, 63, 64, 96, 97, 106, 109, 131, 133, 140, 156, 157
  CO chondrites 16, 19, 49, 63, 81, 98, 99, 157
  EK1-4-1 102
  Fo-bearing inclusions 107
  FUN inclusions 101, 157
  HAL 101, 102
  plagioclase-rich inclusions 107
  poikylitic–olivine inclusions 79
  rare earth elements 107, 128
  spinel–hibonite–perovskite aggregates 5, 106, 118, 137
  spinel–pyroxene aggregates 106
  TE 101, 102
  Type A 106
  Type B 106
  Type C 106
regolith 41, 45, 137, 139
  asteroid 4, 10, 11, 12, 13, 21–35, 39–47, 66, 69, 90, 112, 114, 116, 119, 122, 123, 129–142, 148, 151–162
  depth 38, 43, 45, 123
  formation of chondrules 117, 134, 141, 152, 157
  processes 6, 7, 8, 9, 13, 15, 20, 29, 30, 31, 32, 33, 34, 35, 40, 41, 43, 44, 45, 47, 50, 51, 55, 58, 59, 60, 63, 64, 82, 88, 89, 90, 91, 93, 95, 101, 107, 110–112, 118–120, 123–127, 140–143, 152, 157, 158, 159, 160
  properties 3, 6, 8, 11, 14, 15, 21, 27, 29, 30–35, 43, 45, 49, 50–82, 85, 86, 88, 89, 90, 92–98, 101–114, 122, 126–128, 131, 133, 134, 141, 152, 156, 159
  strength 12, 22, 43, 45, 61, 114, 116
relic grains 92, 93
resonances 24, 25, 113, 117, 138
Reynold's number 148
rims 1, 73, 75, 88, 89, 94, 95, 96, 97, 98, 101, 107, 110, 114, 130
  accretionary 64, 98, 110
  opaque 5, 27, 75, 94, 95, 114
  opaque-rich 95
  significance 10, 70, 107, 121
  thickness 44, 45, 95, 96, 98
rubble pile structures 39, 63
Russell, H. N. 6

S asteroid paradox 28
S asteroids 21, 22, 25, 26, 28, 29, 69
  abundance 7, 8, 9, 14, 15, 27, 33, 39, 51, 53, 54, 55, 57, 58, 60, 68, 73, 80, 82, 83, 88, 89, 90, 104, 109, 127–129, 140, 145, 148, 152, 157
Saturn 24, 25, 125, 137
scatter ellipse 143
shock 10, 30, 32, 34, 43, 58, 61, 67, 75, 111, 113, 117, 118, 119, 122, 123, 138, 152
shock waves 113, 117

siderophile elements 7
  metal–silicate fractionation
solar wind 29, 30, 31, 40, 60
solar wind gases 30, 31
solar–electric propulsion 160, 161
Sorby, Henry Clifton 4
space weathering 26, 30, 31, 32, 132, 133, 159
spacecraft 10, 11, 40, 41, 43–45, 130, 137, 138, 160, 161, 162
  Deep Space 1, 43
  Galileo 10, 40, 41, 45
  Genesis 11, 59
  Hayabusa 160, 161
  Hera 161
  Mariner 9, 40
  Mars Global Surveyor 40
  MUSES C 160
  NEAR Shoemaker 11, 40, 41, 42, 43
  Phobos 2, 40
  Stardust 11, 43
  Viking 1, 40
spectra 21, 25–29, 31, 40, 112, 132, 159, 160
spherules 47, 90, 109, 110, 114, 120–122, 130, 140, 152, 155
  ablation 103, 111, 114
  crystal-bearing 90, 109
  crystallization 35, 37, 60, 85, 90, 92, 110, 121
  glass 18, 31, 60, 64, 75, 77, 110, 112, 120, 121, 136, 140
  impact-melt 47, 109, 110
  volcanic 20, 60, 109, 112, 121, 140, 143, 152, 153
Stoke's drag *see* drag, Stokes
Story-Maskelyne, Nevil 4
suevite 110
Sun 6, 7, 9, 11, 24, 25, 55, 61, 65, 69, 70, 112, 113, 114, 116, 125, 126, 127, 133, 134, 137, 142, 146
  composition 6, 7, 11, 13–15, 17, 19, 21, 22, 31, 35–37, 39, 44, 45, 49–63, 73, 77, 79–109, 114, 120, 125–127, 130, 133–135, 140, 151, 152, 156, 159
  τ-Tauri phase 61, 114
SUNOCONS 118
supercooling 53, 85
supersaturation 85, 115, 134, 144

taenite τ-Tauri phase
temperature 17, 32, 34, 35, 36, 37, 39, 51, 53, 55, 56, 57, 61, 62, 63, 66, 67, 85, 89, 90, 92, 107, 114–116, 118, 119, 126–128, 133, 144, 145, 146, 157
  condensation 31, 53, 55–57, 87, 88, 90, 106, 113–115, 122, 123, 127, 129, 134, 142, 144, 145
  equilibrium 51, 53, 55, 57, 83, 92, 106, 115, 123, 125, 127, 129, 135
  metamorphic 19, 35, 37, 61, 63, 93, 158
TFL line 158
thermal history 35, 37, 63
  impact 10, 11, 15, 20, 22, 25, 31, 35–47, 60, 61, 88, 109, 111–113, 117–123, 129–140, 143, 148, 151–153, 155–159

thermoluminescence 17, 19, 31, 69, 80
turbulent shear 117

undulose extinction 121
Urey, Harold 6
Urey–Craig diagram 8, 49

velocity 46, 47, 115, 118, 119, 122, 123, 130, 138, 142, 143, 149
  impact 10, 11, 15, 20, 22, 25, 31, 35, 37, 38, 39, 40, 41, 43, 44, 45, 46, 47, 60, 61, 88, 109, 119, 111, 112, 113, 117, 119, 120, 122, 123, 129, 130, 133, 134, 135, 137, 138–140, 143, 148, 151–153, 155–159

volcanic spherules 109
volcanism 66, 112, 119, 122, 152, 153
  early ideas

water 12, 14, 15, 19, 21, 22, 23, 31, 57, 64, 94, 97, 104, 119, 121, 122, 139, 147, 148, 151, 152, 155–157
  asteroids 4, 10, 11, 13, 21–47, 69, 129–132, 137–140, 142, 148, 152–156, 159, 160, 161, 162
  chondrites 6–39, 49–86, 93, 94, 96, 97, 98, 99, 102–109, 111, 119, 121, 122, 126–141, 145–159

Yarkovsky effect 25